Columbus

Columbus

Felipe Fernández-Armesto

Oxford   New York

OXFORD UNIVERSITY PRESS

1991

Oxford University Press, Walton Street, Oxford OX2 6DP
Oxford New York Toronto
Delhi Bombay Calcutta Madras Karachi
Petaling Jaya Singapore Hong Kong Tokyo
Nairobi Dar es Salaam Cape Town
Melbourne Auckland
and associated companies in
Berlin Ibadan

Oxford is a trade mark of Oxford University Press

British Library Cataloguing in Publication Data
Data available

Library of Congress Cataloging in Publication Data
Fernández-Armesto, Felipe.
Columbus / Felipe Fernández-Armesto.
p.   cm.
Includes bibliographical references and index.
1. Columbus, Christopher.   2. Explorers—America—Biography.
3. Explorers—Spain—Biography.   4. America—Discovery and
exploration—Spanish.   I. Title.
E111.F35   1991   970.01'5—dc20   90-25592
ISBN 0-19-215898-8

Typeset by Rowland Phototypesetting Ltd
Bury St Edmunds, Suffolk

Such men as thou, who wade across the world
To make an epoch, bless, confuse, appal,
Are in the elemental ages' chart
—Like meanest insects on obscurest leaves—
But incidents and grooves of Earth's unfolding.

(Thomas Hardy, *The Dynasts*)

They all laughed at Christopher Columbus
When he said the world was round.
. . . .
But—ho, ho, ho!—who's had the last laugh now?

(Jerome Kern, *Roberta*)

If it strikes often enough, a drop of water can wear a hole in a
stone.

(Columbus, *La historia del viaje que hizo la tercera vez*)

PREFACE

CONSIDERED from one point of view, Columbus was a crank. Even in his own lifetime he had a cranky reputation. His patrons smiled at his scheme for a crusade and courtiers treated it as a joke.[1] On his first crossing of the Atlantic, mutineers plotted to pitch him overboard during his abstracted machinations with new-fangled and unwieldy navigational instruments.[2] He claimed to hear celestial voices.[3] He embarrassed the court of the Spanish monarchs by appearing provocatively attired in public, once in chains and regularly in a Franciscan habit.[4]

These eccentricities are easy to excuse or even to applaud as such imps as often attend genius. They have had, however, one regrettable effect. Columbus has attracted cranks, as crag calls forth to crag; and if one of the many committees convened to honour the fifth centenary of the discovery of America were to offer a prize for the silliest theory about him, the competition would be keenly contested. Readers wanting to know about Columbus might be almost as badly misled by the many well-meaning amateurs who have been induced by his presumed importance to write up his life: most books about Columbus have been biographies, which even at their best can seem to abstract their protagonist from his proper context. Overwhelmingly the effect has been to project, into popular books, versions of a Columbus who was 'ahead of his time'—a Columbus inaccessible to an imagination disciplined by respect for the sources and by knowledge of the period. If scholarly biographies so far, with few exceptions, have not yielded any more convincing general impression of Columbus, misleading influence from sixteenth-century writers, loosely treated as primary sources, is probably to blame.[5] For five hundred years, Columbus historiography has been afloat without heeding the need for a good long spell in dry dock. Like a well-barnacled bottom, it needs a vigorous scrape to get rid of the glutinous concretion of errors and false impressions. When restored to deep water, it has to be steered cautiously to elude the cranky theories and undisciplined speculations alike. In the Sea of Darkness, Siren voices rise on every side.

This book has been written in the belief that readers want unadorned facts about Columbus, as far as these can be elicited. I have tried to say nothing which cannot be verified—or in some cases reasonably inferred

—from unimpeachable sources. Narratives of the sixteenth century have been excluded, except where they can be shown to reflect sources otherwise lost, or for the occasional *aperçu* which seems useful to me and which has been clearly signposted, with a warning to the reader, in the text or endnotes. Even accounts written soon after Columbus's death by privileged observers have been sparingly used, subject to corroboration. Columbus's own narratives, which can hardly be forgone, have been handled tentatively and scrutinized carefully for the promotional or exculpatory purposes that distorted almost every thought Columbus ever confided to paper. One result of my reliance on Columbus's own writings, and of my sceptical treatment of them, is that much of this book is not so much about what happened to Columbus as about what was going on in his mind, which—surprisingly, perhaps—is easier to know.

The Columbus who emerges may not be much more objective than any other, as his image bounces flickeringly between the reader's retina and my own. The Columbus I detect—the socially ambitious, socially awkward parvenu; the autodidact, intellectually aggressive but easily cowed; the embittered escapee from distressing realities; the adventurer inhibited by fear of failure—is, I believe, consistent with the evidence; but it would no doubt be possible to reconstruct the image, from the same evidence, in other ways. Other students have imagined him essentially as a practical tarpaulin, or a ruthless materialist, or a mystic seer, or an embodiment of bourgeois capitalism; the springs of his motivation have been perceived in an evangelical impulse, or in some more generalized religious conviction, or in crusading zeal, or in scientific curiosity, or in esoteric or even 'secret' knowledge, or in greed. I find these versions unconvincing, but I have not written in order to advance my view at their expense—only to satisfy readers who want to make their own choices from within the range of genuine possibilities.

There are, however, three traditions of Columbus historiography which I actively defy. The first is the mystifying tradition, concerned to reveal allegedly cryptic truths which the evidence cannot disclose. Works of this type argue either that Columbus was not what he seemed, or that his plan for an Atlantic crossing concealed some secret objective. For instance, the rationally unchallengeable evidence of Columbus's Genoese provenance has not prevented mystifiers from concocting a Portuguese, Castilian, Catalan, Majorcan, Galician, or Ibizan Columbus, sometimes with the aid of forged documents.[6] At a further level of mystification, a persistent tradition has insisted on a Jewish

Columbus. His own attitude to Jews was not free of ambivalence: at one level he treated them with respect and professed, for instance, that, like Moors and pagans, they could be accessible to the operations of the Holy Spirit; at another level he shared the typical prejudices of his day, condemning the Jews as a 'reprobate' source of heretical depravity and accusing his enemies of the taint of Jewish provenance.[7] The theory that he was of Jewish faith or origins himself can only be advocated *ex silentio*, in default—and sometimes defiance—of evidence.[8]

Believers in Columbus's 'secrets' thrive on lack of evidence, because, like every irrational faith, theirs is fed on indifference to proof. Thus otherwise creditable scholars have argued, for instance, that all the evidence which proves that Columbus sailed in 1492 on a mission to Asia should be 'decoded' to demonstrate the opposite; or that his plan can be explained only by access to secret foreknowledge, transmitted by an 'unknown pilot', or by means of a fortuitous pre-discovery of America by Columbus himself, or even as the result of a chance encounter with American Indians.[9] Readers of this book can rely on being spared any such rash speculations.

The second objectionable tradition treats paucity of evidence as a pretext for intuitive guesswork. Imaginative reconstructions of what Columbus 'must' have been thinking or doing at moments when the sources are silent or ignored are made the basis for vacuous conclusions. On the strength of such musings, in highly popular books, Columbus has been credited with a strenuous love-life, with visionary glimpses of America from Iceland or Porto Santo, with undocumented visitations by his 'voices', and with a plan to conceal his presumed Hebraic ancestry.[10] Sometimes the method is defended by frank contempt for the essential resources of historical enquiry, by an appeal to 'leave the dusty documents on the shelf and come back to the flesh and the spirit' or to speculation licensed on the grounds that 'there are no documents, only the real lives of these men and women, whose blood coursed through their veins as does ours through our own'.[11] Yet, even if one were disposed to admit this obviously fallacious reasoning, the premiss on which it is based is false. We are extremely well informed about Columbus. No contemporary of humble origins or maritime vocation has left so many traces in the records, or so much writing of his own.

The last hazard I have tried to avoid is that of subscribing to a legend of the explorer's own making. The picture transmitted by the historical tradition of a uniquely single-minded figure is false, I am sure. Though Columbus could be obsessively pig-headed, his self-image, as I try to show in this book, was dappled by doubts. His sense of divine purpose grew gradually and fitfully and was born and nourished in adversity. His geographical ideas took shape slowly and were highly volatile in the early stages. His mental development proceeded by fits and starts and led at different times in different directions. The contrary view—that his ideas came suddenly, as if by revelation or 'secret' disclosure, or were sustained consistently, in defiance of contemporary derision, with an inflexible sense of purpose—goes back to a 'promotional' image which Columbus projected in his own writings in the latter part of his life. His aim was not only to dramatize his story and to emphasize the unique basis of his claims to material rewards but also to support a broader picture of himself as a providential agent. He was, he professed, divinely elected to execute a part of God's plan for mankind, by making the gospel audible in unevangelized parts of the earth. That tendentious reading of his own life was adopted by the authors of the detailed sixteenth-century narratives that have influenced all subsequent writers. Bartolomé de Las Casas, whose work has been fundamental to all modern studies of Columbus, accepted Columbus's self-evaluation as a divine messenger because he shared a providential vision of history and wrote to justify and celebrate an apostolate among the Indians in which he personally played no mean part; the next most influential narrative, the *Historie dell'Ammiraglio*, reflects much of the same view, either because it was derived from Las Casas's work, or perhaps because it was genuinely the work of Columbus's son, to whom it is attributed.[12] Although few modern historians admit to a providential conception of history, almost all have accepted a secularized version of the legend, generally with misleading results. Some wild conclusions have been based, for instance, on the myth of Columbus's 'certainty', which goes back to Las Casas's vivid image: 'so sure was he of what he would discover, that it was as if he kept it in a chamber locked with his own key.'[13]

Columbus is seen best—understood, that is, most fully—in the contexts in which he belonged: the Genoese world of the late fifteenth century; the partly Genoese Lisbon and Andalusia to which he moved at a critical period of his career; the court of the Spanish monarchs, which was effectively his base of operations in the second half of his life; the mapping and exploration of the Atlantic in his day; the world of geographical speculation by which he was surrounded; and, in a remoter background, the slow shift of the centre of gravity of Western civilization from the Mediterranean to the Atlantic, to which he made such an important contribution. I have tried to sketch these in briefly. Historians nowadays should make only calculated claims on their readers' time, and the most important purpose of this book is to cover the essentials, decently but with becoming brevity.

Almost everything I know about Columbus has been learnt in ten years of teaching for papers based partly on his writings in the Honour Schools of Modern History and Medieval and Modern Languages in the University of Oxford. Among colleagues and pupils, I am particularly indebted to Roger Highfield, Penry Williams, John Hopewell, and Alina Gruszka. My errors, like those of Columbus, spring from indifference to advice.[14]

F.F.-A.

Partney House, Lincolnshire
July 1990

CONTENTS

CONTENTS

LIST OF ILLUSTRATIONS

LIST OF MAPS

CHRONOLOGY

1429   Columbus's future father, Domenico Colombo, apprenticed as a weaver.

c.1445   Domenico Colombo marries Susanna Fontanarossa.

c.1451   Columbus born in or near Genoa.

1472   Columbus associated with his father in family weaving business.

c.1476   Columbus moves to Lisbon.

1477?   Makes voyage to Iceland, perhaps via England and Ireland.

1478   Sugar-buying trip on behalf of Centurione firm recorded to Madeira.

c.1479?   Marries Felipa Perestrello e Moniz.

1480   Birth of his elder son, Diego.

1482–5?   Voyage(s?) to São Jorge da Mina.

1484?   First approaches King João II of Portugal with a project for an Atlantic crossing.

1485?   Transfers his quest for patronage to Castile.

1486   May: Has audience with Ferdinand and Isabella in Cordova.

1488   November: Birth of his son Fernando to Beatriz Enríquez.

December: Columbus apparently in Lisbon again.

1489/90?   Columbus's brother Bartolomé offers the project in France and England.

1489   May: Columbus at the court of Ferdinand and Isabella.

1492   January: Financial consortium directed by Luis de Santángel decides to back Columbus.

17 April: Columbus contracts with the Castilian monarchs to find 'islands and mainland in the Ocean Sea'.

23 June: Columbus obtains ships and recruits men in Palos.

3 August: Departs Palos in an attempt to cross the Atlantic.

12 August: Arrives in Canaries; makes modifications and repairs to ships.

6 September: Departs San Sebastián de la Gomera.

12 October: Arrives in Bahamas, at an unidentifiable island.

28 October: Discovers Cuba.

6 December: Discovers Hispaniola.

24 December: Flagship (*Santa María*) grounded and abandoned.

25 December: Fort founded at Navidad, Hispaniola.

1493   16 January: Columbus departs Hispaniola.

14 February: Columbus has his first recorded experience of his celestial 'voice'.

17 February: Sights Santa Maria, Azores; disembarks 18 February.

4 March: Arrives Lisbon; departs 11 March.

15 March: Arrives Palos.

April: Columbus reports to Ferdinand and Isabella in person at Barcelona.

3/4 May: Pope Alexander VI issues Bulls *Inter Caetera*, granting sovereignty over Columbus's discoveries to Castile.

25 September: Columbus departs on his second Atlantic crossing.

13 October: Leaves Hierro, bound across the Atlantic.

3 November: Makes landfall, Dominica.

18 November: After a series of discoveries in the Lesser Antilles, Columbus discovers Puerto Rico.

23 November: Arrives Hispaniola.

28 November: Arrives at fort at Navidad, finds garrison massacred.

1494   24 April–29 September: Columbus absent from Hispaniola, exploring Cuba and Jamaica.

29 September: Columbus reunited with his brother Bartolomé in Hispaniola.

1495   March: Columbus begins a series of campaigns, lasting more than a year, to subdue the interior of Hispaniola.

October: Juan Aguado arrives to conduct judicial investigation into Columbus's discharge of his duties as governor.

1496   10 March: Columbus departs Hispaniola, bound for Spain, via Marigalante and Guadelupe; arrives Cadiz 11 June.

1497   23 April: Ferdinand and Isabella issue first instructions for Columbus's third crossing.

Summer: Columbus spends some time at the friary of La Mejorada.

1498   30 May: Departs Sanlúcar de Barrameda on third Atlantic crossing, via the Cape Verde Islands.

1 July: Arrives São Tiago; departs 4 July.

31 July: Trinidad sighted.

2–13 August: Explores coast of American mainland along the Paria peninsula.

14/15 August: Records opinion that he has discovered 'a very great continent, which until today has been unknown'.

19 August: Arrives at Hispaniola; finds Roldán's rebellion in progress.

1499   September: End of Roldán's rebellion. Alonso de Hojeda lands in Hispaniola.

25 December: Columbus visited by his celestial voice.

1500   June: Vicente Yáñez Pinzón lands in Hispaniola.

August: Columbus suppresses another rebellion under Adrián de Moxica; Francisco de Bobadilla arrives to conduct judicial investigation.

September: Columbus arrested.

October: Columbus shipped back to Spain in chains.

16 December: Columbus presents himself to Ferdinand and Isabella, and is well received.

1501 February: Columbus's surviving correspondence with Gaspar de Gorricio begins; he is perhaps at work on the *Book of Prophecies*.

13 September: Nicolás de Ovando appointed governor of Hispaniola.

1502 13 February: Ovando departs.

14 March: Ferdinand and Isabella grant Columbus permission to make a fourth crossing.

3 April: Departs from Seville on fourth and final crossing; delayed by weather, leaves Cadiz 11 May.

20 May: Arrives Gran Canaria; departs 25 May.

15 June: Arrives Martinique; 29 June arrives off Hispaniola, where, consistently with the monarchs' orders, he is denied shelter in Santo Domingo from an approaching storm.

30 June: The storm spares Columbus's fleet and treasure, but most of the homeward-bound fleet is destroyed and Francisco de Bobadilla killed.

14 July: Leaves Yáquimo (south coast of Hispaniola).

30 July: Arrives Bonacca; begins voyage along central American isthmus.

20 October: Discovers gold-rich province of Veragua.

2 November: Discovers harbour of Porto Bello.

5 November: Forced back to Veragua by bad weather.

10 November: Discovers Nombre de Dios Bay.

1503 6 January: Anchors at Rio Belén.

6 April: Visited by his celestial voice.

16 April: Escapes from almost enclosed river-mouth at Belén with loss of one ship; follows coast in badly worm-ridden ships in the hope of reaching the meridian of Hispaniola.

1 May: Heads north into open sea; after being blown as far as Cuba, reaches Jamaica on 23 June.

25 June: Abandons ships in St Ann's Bay, Jamaica.

July–August: While Columbus's force is marooned on Jamaica, Diego Méndez reaches Santo Domingo by canoe and overland to seek help, which governor Ovando refuses to provide. Mutiny of the Porras brothers.

1504 29 February: By predicting an eclipse, Columbus intimidates the natives of Jamaica into keeping his men supplied.

March: Diego de Escobar visits Columbus's camp from Hispaniola.

19 May: Porras mutiny suppressed.

June: Relief ship organized by Diego Méndez rescues Columbus and his men; they arrive at Yáquimo on 3 August.

12 September: Columbus sails for Spain.

7 November: Arrives at Sanlúcar.

26 November: Death of Queen Isabella.

December: Diego and Bartolomé Colón at Ferdinand's court.

1505   May: Columbus, recovering slightly from cripplingly poor health, journeys to court.

Early summer: Has unsatisfactory audience with the King in Segovia.

25 August: Adds codicil to his will.

1506   April: The new King and Queen, Felipe and Juana, arrive in Spain. Columbus writes them his last known letter.

20 May: Columbus dies in Valladolid.

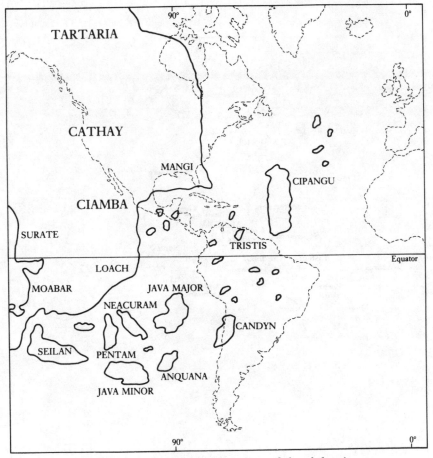

Martin Behaim's Conception of the Atlantic

Columbus in the Old World

ANNOTATIONS

AZORES: Flotsam said to have been collected here in an early biography.

BRISTOL: Visit inferred from the context of his Iceland voyage and his description of the height of tides.

CANARIES: First-hand knowledge before 1492 inferred from the general context of Columbus's Atlantic experience and from his choice of the islands as the starting-point for his Atlantic crossing. His references to the topography and inhabitants of the islands show at least superficial acquaintance.

CHIOS: Visits here in connection with the mastic trade alluded to (and once directly recalled) in Columbus's writings.

CORDOVA: First interview with Ferdinand and Isabella, May 1486.

ELMINA: In annotations to his books, Columbus claims to have been here on a voyage which can be assigned to between 1482 and 1485. His references to the area and his use of terms peculiar to the Portuguese argot of the coast are corroborative. His claim to have made readings of latitude on the journey should be judged in the light of the attempt by the Portuguese court astonomer, José Vizinho, to use the route to refine latitude-finding techniques.

FLORENCE: Linked to Columbus by a surviving copy, in Columbus's hand, of a letter of 1474, in which Paolo del Pozzo Toscanelli confides to a Portuguese correspondent a plan for crossing the Atlantic to Asia. According to the probably interdependent reports of his early editor and biographer, Columbus was in touch with Toscanelli before the latter's death in 1482 and received a letter which has survived only in purported copies, enclosing the 1474 document and a speculative map of the route.

FUENTERRABIA AND COLLIOURE: Cited by Columbus in 1493 as the extremities of the coastline of Spain.

GALWAY: In an undated marginal annotation Columbus records having seen castaways, whom he presumed to be of Asiatic provenance at this routine port of call of northern trade.

GENOA: Birthplace. Corroborated by many references to his Genoese provenance and a document in which his cousins declare the relationship.

GRANADA: Columbus receives royal permission for his Atlantic voyage, January, 1492.

ICELAND: In 1495 Columbus claimed to recall a voyage to 'Thule and a hundred leagues beyond' of eighteen years before.

JERUSALEM: Columbus's crusade proposed before 1492.

LISBON: Presence here, documented at intervals, can be presumed to have begun in the mid-1470s and certainly before 1477.

MADEIRA: Voyage as a sugar-buyer recorded in a document of 1478.

NUREMBERG: Cosmographers here, perhaps independently, evolved a scheme similar to Toscanelli's by 1492, when Martin Behaim made his globe with its depiction of a navigably narrow Atlantic

PALOS: Traditionally from 1485, certainly from 1491, Columbus cultivated links with the Franciscans of La Rábida. Shipping and manpower were recruited here with the help of the leading sea-going and ship-owning family, the Pinzón.

PORTO SANTO: Home of Columbus's wife, Dona Felipa Moniz, whom he married in or about 1479.

SALAMANCA: Visits 1486–91 widely presumed but unverified.

SAVONA: Site for a time, including 1472, of the family home and business, which may have involved trading in wool and cloth.

SEVILLE, CADIZ, BARCELONA, GULF OF NARBONNE, MARSEILLES, NAPLES: Places recalled by Columbus in a letter of 1502 as halts on a cabotage route.

TUNIS: In a letter of 1495 Columbus claims to have been on an expedition here, presumably in the early 1470s, from Marseilles.

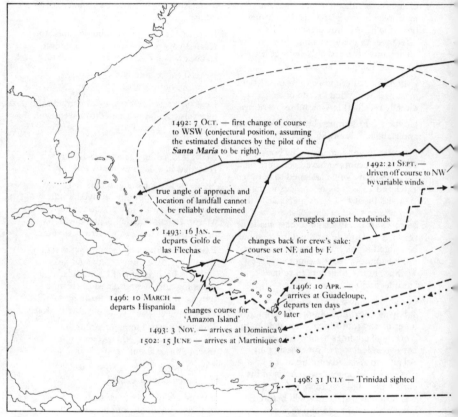

1492: 7 OCT. — first change of course
to WSW (conjectural position, assuming
the estimated distances by the pilot of the
Santa María to be right).

1492: 21 SEPT. —
driven off course to NW
by variable winds

true angle of approach and
location of landfall cannot
be reliably determined

struggles against headwinds

1493: 16 JAN. —
departs Golfo de
las Flechas

changes back for crew's sake:
course set NE and by E.

1496: 10 APR. —
arrives at Guadeloupe,
departs ten days
later

1496: 10 MARCH —
departs Hispaniola

changes course for
'Amazon Island'

1493: 3 NOV. — arrives at Dominica
1502: 15 JUNE — arrives at Martinique

1498: 31 JULY — Trinidad sighted

Columbus's Routes across the Atlantic

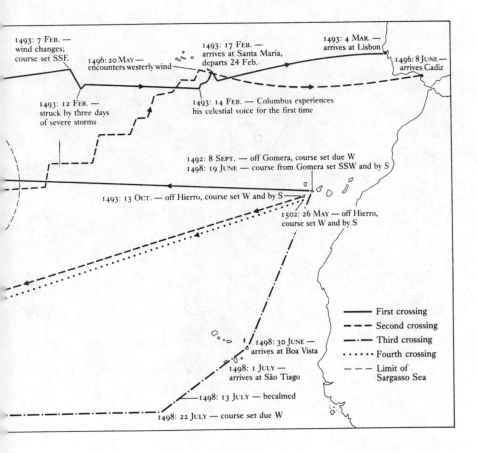

1493: 7 FEB. — wind changes; course set SSE.

1496: 20 MAY — encounters westerly wind

1493: 17 FEB. — arrives at Santa Maria, departs 24 Feb.

1493: 4 MAR. — arrives at Lisbon

1496: 8 JUNE — arrives Cadiz

1493: 12 FEB. — struck by three days of severe storms

1493: 14 FEB. — Columbus experiences his celestial voice for the first time

1492: 8 SEPT. — off Gomera, course set due W
1498: 19 JUNE — course from Gomera set SSW and by S

1493: 13 OCT. — off Hierro, course set W and by S

1502: 26 MAY — off Hierro, course set W and by S

1498: 30 JUNE — arrives at Boa Vista

1498: 1 JULY — arrives at São Tiago

1498: 13 JULY — becalmed

1498: 22 JULY — course set due W

——— First crossing
– – – Second crossing
–·–·– Third crossing
········· Fourth crossing
– – – Limit of Sargasso Sea

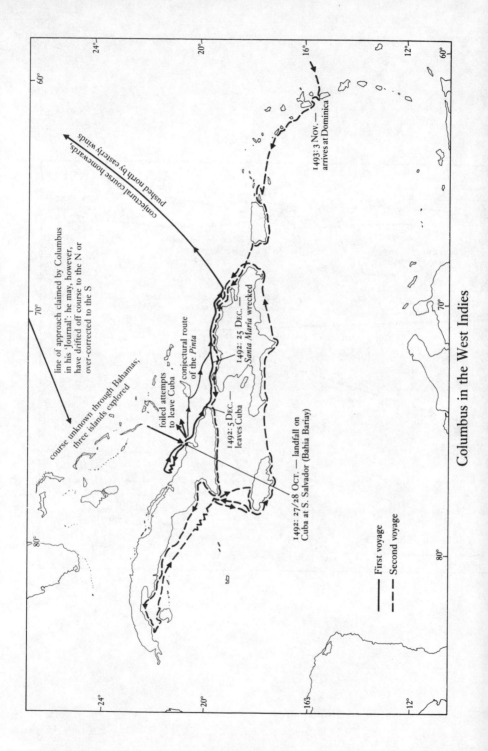

conjectural course homewards, pushed north by easterly winds

line of approach claimed by Columbus in his 'Journal'; he may, however, have drifted off course to the N or over-corrected to the S

course unknown through Bahamas; three islands explored

foiled attempts to leave Cuba

conjectural route of the *Pinta*

1492: 25 DEC. — *Santa Maria* wrecked

1492: 5 DEC. — leaves Cuba

1493: 3 NOV. — arrives at Dominica

1492: 27/28 OCT. — landfall on Cuba at S. Salvador (Bahia Bariay)

—— First voyage

– – – Second voyage

Columbus in the West Indies

1502: 30 JULY — GUANACA (Guanaja, Bonacca).
Columbus encounters trading canoe

PTA CAXINAS (C. Honduras). *Encounters Indians in cotton armour 'of the Kingdom of Maya'. Voyage against the wind begins*

COSTA DE LAS OREJAS

Takes possession for Ferdinand and Isabella

1502: 14 AUG. —
R. DE POSESIÓN
(R. Romano).

'Other tempests have I seen but none so long or bad'

Adverse wind ceases
1502: 14 SEPT. —
C. GRACIAS A DÍOS.

↧ Recorded anchorages

Names used by Columbus in small bold capitals, ie PTO GORDO

R. Grande

2 men lost in foundered boat
1502: 16 SEPT. —
R. DE LOS DESASTRES.

LIMONES (Pearl Cays)

C. DE ROJAS
(Pta del Mono)

S. JUAN DEL NORTE

Indians willing to trade but have no gold
LIMÓN

1502: 5 OCT. —
BOCA DEL DRAGÓN
ZOROBARÓ
(I. Colón)

CARIAI

CIAMBA

GUAIGA

VERAGUA

1502: 17 OCT. —
ESCUDO DE VERAGUA

1503: 6 JAN.–16 APR. —
S. MARÍA DE BELÉN

1502: 25 DEC. — PTO GORDO

1502: 2 NOV. — PTO BELLO

1502: 10–23 NOV. —
BASTIMENTOS (Nombre de Dios)

1502: 26 NOV. – 5 DEC. —
PTO RETRETE (Escribanos?).

Columbus rejects chief's gift of two young virgins, witnesses battle between a boar and a spider monkey

1502: 25 SEPT.–5 OCT. —
LA HUERTA

First signs of gold detected

Indians certify proximity of 'strait'

1502: 6–16 OCT. —
ALBUREMA (L. Chiriqui).

6 APR. —
Experiences celestial 'voice'

Turns back for Veragua; reaches Belén after a month through terrible weather

From about here, driven by violent storm as far as Pto Bello

1503: 1 MAY —
C. MARMOREO?
Turns north in attempt to reach Hispaniola

Columbus's Voyage from Honduras to Darién 1502–3

I

A Man 'Raised from Nought'

FROM GENOA TO THE ATLANTIC,
c.1450–c.1480

COLUMBUS began his career as an escapee from a clannish family and a humble home. It is as certain as knowledge can be from so long ago that the Cristóbal Colón who sailed across the Atlantic in 1492 was the same Cristoforo Colombo who was born in or near Genoa, to a weaver named Domenico, probably a little over forty years before. The evidence consists not only in Columbus's own frequent, apparently heartfelt assertions of his Genoese provenance, but also in a document of unimpeachable authenticity, in which some of his Genoese kinsfolk declared their intention of sending to Spain, after his rise to fame, to seek his patronage.[1] This fact helps to convey a sense of the social trajectory of Columbus's life: the restrictive circle into which he was born and the clinging brood which surrounded him; the escape into worldly success; the clustering of kinsmen around the fortunate *arriviste*; the role of family provider to which he was committed by his hard-won place in the acceptance world. At the end of his life, Columbus was distributing titles of honour and (largely imaginary) wealth to his surviving brothers and their heirs. One thinks of Napoleon—another marginal character from an intensely 'tribal' Italian background—turning his impoverished siblings into sovereigns.

Though not much is known of Columbus's early domestic circumstances, it is evident that he was ashamed of them. He was evasive about his origins. An early biography attributed to his younger son suggests that this was out of modesty. He preferred, his biographer alleges, to rise by merit, rather than relying on his illustrious ancestry to define his place in the world.[2] A concept typical of Renaissance moral philosophy underlies the argument: nobility consists not in ancient lineage but in personal virtue. This, however, like many Renaissance concepts, was one which Columbus did not fully share. He would have vaunted an ancient lineage, had he possessed such a thing. His hankering after illustrious forebears can be detected in the shadowy claim that 'I am not the first admiral of my line'. He was being more frank when he admitted that his patrons had 'raised me up from nought'.[3]

While asserting an admiral-progenitor, he suppressed all mention of his weaver-father. His mother Susanna, a weaver's daughter, was smothered by the same silence, as was his sister Bianchinetta, who married a cheesewright. For the brothers who survived to manhood, however—Bartolomeo and Giacomo, known exclusively according to Castilian orthography as Bartolomé and Diego—Columbus showed due family feeling: 'ties of blood and great love', as he said.[4] Bartolomé was the companion and surrogate of his long years spent supplicating for patronage at the Western courts of Latin Christendom, and the right-hand man of his attempts to establish a colony in the New World. Diego accompanied Columbus's second Atlantic crossing and continued to command his affection and attract his patronage. 'I never had a better friend,' Columbus recalled towards the end of his life, 'in fair weather or foul, than my brothers.' From a maker of fair-weather friends, many of whom deserted him, this was not, perhaps, saying very much. The mutual loyalty of Columbus and his brothers during his period in Spanish service gave him the comfort of family solidarity, in a foreign land and hostile company, but provoked resentment from excluded subordinates on voyages and in colonial commands. Cousins of different generations, Giovanni Antonio and Andrea, served on the third and fourth Atlantic crossings respectively, stoking the disenchantment which some of Columbus's other followers felt for the whole brood.[5]

Domenico Colombo may not have been an easy father to be proud of. If, as is likely, he can be identified with the weaver of the same name who is recorded doubling as a tavern-landlord in Savona, he can be presumed to have made some effort to improve his fortune; but in 1473 he was in need of ready money, liquidating movable assets, and ten years later was under pressure from his creditors to sell his house. A less certain identification is with Domenico Colombo, keeper of a city gate in 1447 and 1450; the exercise of even so modest an office of profit could not have been obtained without the patronage of one of the factions active in the turbulent history of Genoese politics, but this line of enquiry, though intriguing, is too shadowy to pursue with profit.[6]

'Columbus the pauper'—as an early commentator called him[7]—would not have suppressed mention of such undistinguished parents for reasons of modesty. The obscurity of his origins is a sufficient explanation of his reticence. Nor was Columbus ever a man to do anything for modesty's sake. Even the humility he affected in later life, when he went about in a rough habit,[8] was of a showy, exhibitionistic kind. He claimed to have been divinely inspired—which is a curiously egotistical form of self-effacement. The role of great nobleman and 'captain of conquests', devised for himself in the latter years of his life, was, for an uneducated man, uncannily well scripted and the script impressively well learned.[9]

From one point of view, the most consistent single purpose to which his own life was dedicated was the desire to found a noble dynasty of his own. The priorities he claimed—the service of God and of the monarchs of Spain, the advancement of science—appear by comparison subordinate or ancillary: threads in the cloak of self-aggrandizement which the ex-weaver cut to suit himself. Grumbling shipmates on his first Atlantic crossing got his priorities right: all he wanted, they complained, was 'to be a great lord' and he was willing to risk his life and theirs for it.[10] On his last voyage he disclaimed any wish for 'status and wealth' but admitted, by implication, that they had been his objectives until then.[11]

The Church and war were the main channels of upward mobility in Columbus's day. His youngest brother Diego may have decided, at an unknown date before 1498, to pursue a clerical career;[12] but although Columbus developed some unusually strong religious sensibilities late in life, he is not known to have evinced any similar vocation when young. Just as he came to see himself as a quasi-sacerdotal figure—clad in Franciscan garb, bearing the light of the gospel to pagans—in the last decade of his life, so he also, at about the same time, adopted a soldierly self-perception as a 'captain sent from Spain to conquer a numerous and warlike people.'[13] Both these affectations, however, were late additions to his mental baggage. Spirituality was embraced, as we shall see, as a refuge from adversity; the soldierly role was alleged, during his period of disgrace in 1500, to mask his deficiencies as an administrator. In a letter written in 1495 he implied, almost certainly misleadingly, that he had an independent command at sea in his youth in the wars between the Angevin and Aragonese dynasties for control of the Kingdom of Naples.[14] With this exception, there is no hint that Columbus ever had the chance of taking a warlike route to self-advancement.

But the fifteenth century also offered the socially ambitious a sea-way to elevated goals, usually across island stepping-stones. One of the most popular new books in the Spain of Ferdinand and Isabella was the one described in *Don Quixote* as 'the best in the world'—the *Tirant lo blanc* of Joan Martorell, an extravagant chivalric romance, in which one of the characters is a 'king of the Canary Islands', who launches, with presumably conscious irony on the author's part, an invasion of Europe. The erection of an island-kingdom is a common dénouement in works of the genre: in *Don Quixote* the tradition is derided through Sancho Panza's aspiration to rule an island. When Ferdinand and Isabella added the style of 'King and Queen of the Canary Islands' to the list of titles with which they headed their letters, they were exceeding fiction and making a reality of romance and Columbus played on the same tradition when he addressed them as 'King and Queen of the Islands of the Ocean'. Identical alchemy transformed base men into princes or governors earlier in the century. A cut-throat member of Henry the Navigator's rabble of 'knights and squires' was transmuted into 'Tristram of the Island' by service on Madeira. The Norman adventurer Jean de Béthencourt had himself proclaimed King of the Canary Islands on the streets of Seville. One of the men 'made' by island-escapades in the 'Ocean Sea' (as the Atlantic was then called) was Bartolomeo Perestrello, sent from the entourage of Henry the Navigator to settle and govern Pôrto Santo: marriage to his daughter, at an unknown date between 1477 and 1480, was to be Columbus's first big step towards social cachet.[15]

The world of maritime adventure which Columbus joined was best evoked, perhaps, by the figure of Count Pero Niño, whose chronicle, written by his standard-bearer in the second quarter of the fifteenth century, is a treatise of chivalry as well as an account of campaigns: *El vitorial* celebrates a knight never vanquished in joust or war or love, whose greatest battles were fought at sea; and 'to win a battle is the greatest good and the greatest glory of life'. When the author discourses on the mutability of life, his interlocutors are Fortune and Wind, whose 'mother' is the sea 'and therein is my chief office'. Columbus's younger contemporary, the Portuguese poet Gil Vicente, was able—thanks to the chivalric connotations of the sea—to liken a lovely woman to a ship and a warhorse without incongruity. It was as if romance could be sensed amid the rats and hardtack of shipboard life, or the waves ridden like jennets. There is no evidence that Columbus ever read any of the chivalric literature of the sea, but he moved in a world steeped in it; his life was, in a sense, an embodiment of it; and the islands that decorated the coat of arms he won for himself were an image of it.[16]

For a Genoese boy of modest origins and little education, seafaring was a perfectly natural choice of career. Columbus was frank about his lack of formal schooling—most eloquently so in a retrospect of 1501, written when the climax of his career was past and his health and fortunes were already in decline:

Every sea so far traversed have I sailed. I have conversed and exchanged ideas with learned men, churchmen and laymen, Latins and Greeks, Jews and Moors and many others of other religions. To that wish of mine I found that Our Lord was very favourably disposed, and for it He gave me the spirit of understanding. He endowed me abundantly in seamanship; of astrology He gave me sufficient, and of geometry and arithmetic too, with the wit and craftsmanship to make representations of the globe and draw on them the cities, rivers and mountains, islands and harbours, all in their proper places. Throughout this time I have seen and studied books of every sort—geography, history, chronicles, philosophy and other arts—whereby Our Lord opened my understanding with His manifest hand to the fact that it was practicable to sail from here to the Indies.[17]

Though intended as an explanation of the genesis of his project for an Atlantic crossing, this passage has the effect of describing his own long, slow process of self-education. Like other religious men in a similar position, Columbus ascribed to God what was evidently his own role in acquiring the combination of book-learning and practical sagacity which distinguished him in maturity. From the way he expresses it, it is obvious that little or none of this equipment was acquired in childhood.

He certainly never attended a university: his supposed place among the alumni of Pavia was an invention of an early biographer's.[18] The appraisal of him by a friend—the priest and chronicler of his own times, Andrés Bernáldez—as a man of 'great intellect but little education'[19] was accurate. He had the characteristic intellectual shortcomings of a self-educated man. His mind suffered the defects that a guideless and random absorption of knowledge can impart, like a ship at large upon a starless ocean. He read intently, but not critically; he acquired, over a long time, a mass of information, but was never able to dispose of it to best advantage. He could mimic a variety of styles in a number of languages, but always made silly or risible errors. He would leap—in his attempts at reasoning—to bizarre conclusions, on the flimsiest evidence, which a more balanced preparation might have taught him to eschew. He selected his reading obsessively, choosing whatever supported his own theories, rejecting or distorting whatever would not fit.

In any case, by his own account his youthful travels preceded his self-education analytically as well as in time. 'From a very small age,' he wrote in 1501, 'I went sailing upon the sea, which very occupation inclines all who follow it to wish to learn the secrets of the world.'[20] His navigational experiences, that is, drew him as if blown by a wind and pulled by a current into the ocean of geographical speculation. The quest for seaborne glory is most likely, perhaps, to have come upon Columbus gradually as his seafaring experience increased. It would be rash to suppose that he began his nautical career with any specific ambitions. Apart from his Genoese birth, the only other credible assertion Columbus made about his early life was the claim that he went to sea 'at a very early age'. The date is unknown. The biography ascribed to his son says that it was when the future discoverer was fourteen. In 1492 Columbus dated the same event twenty-three years earlier (if the transcription of the document can be trusted).[21] In 1472 Columbus was still involved, at least part time, in the family weaving business—though this would not preclude sea-journeys made, for instance, in the course of buying wool or selling cloth.[22]

The navigations he can safely be said to have undertaken between the early 1470s and the mid-1480s can be reconstructed from more or less random survivals in the sources (see Map 1). On the face of it they appear to encompass an astonishing range, which took him not only around Genoa's home waters in the Ligurian and Tyrrhenian Seas[23] but also east to the limits of the Mediterranean, in Chios,[24] and west to the remotest points of established navigation in the Atlantic: to Iceland in the north, the Azores in the centre, and the Gulf of Guinea in the south.[25] This impressive record of schooling in seamanship must be treated with caution as it derives almost entirely from Columbus's own testimony. It makes sense, however, against the background of the Genoa of his day—a background which, if examined closely, can make up for the deficiencies of our knowledge of Columbus's early life by conveying a flavour of the world in which he moved.

In his mid-sixteenth-century *Cosmographia*, Sebastian Münster chose to represent Genoa with a Janus-figure brandishing a large key.[26] A more popular medieval legend derived the city's name from the supposed Trojan founder Ianos, but Münster's conceit better represents the character of Genoa as it had come to be defined in the late Middle Ages:

a Janus facing east and west, on the one hand towards the trade of the
Levant, the Black Sea, and the Orient, on the other towards the western
Mediterranean, the Maghrib, and the Iberian peninsula. From the late
thirteenth century, when Genoese ships in large numbers began to cut
through the adverse current that locked Mediterranean shipping within
the Straits of Hercules, that westward gaze was extended ever farther
into the Atlantic. Though punningly thought of as the 'door' (*ianua* in
Latin) to Italy along the Ligurian coast road, Genoa never controlled
access by land around or across the Alps. Her growing maritime power,
however, her sketchy but far-flung 'empire' of merchant colonies along
the Iberian and Maghribi sea-routes to the Atlantic and her hugely
disproportionate stake in Mediterranean–Atlantic trade gave her a
privileged position in the late medieval opening of the Atlantic. The key
Janus holds in Münster's drawing should be thought of as opening the
'door' not of the old Roman road from Gaul to Italy by the coast, but of
the Pillars of Hercules.

The Genoese network of centres of production and exchange was an
empire only in the feeblest of senses: first, because it lacked central
direction from the institutions of the State; secondly, because it contained few sovereign colonies; thirdly, because of the ambivalence of
Genoese traders, whose success owed much to their mutual solidarity
but more to their adaptability and gift for distinguishing and serving
private or family interests rather than those of their nation. Furthermore, Genoese policy had a 'hermit crab' character, content wherever
possible to work within or alongside other states. From Byzantium and
the Khanate of the Golden Horde in the east to Portugal and Castile in
the west, the Genoese accepted the protection of foreign princes; the
result was a form of covert colonialism or surrogate empire-building, in
which, for example, much of the profit from Castilian overseas expansion was drawn, by a flick of the purse-strings, into Genoese hands. A
further effect was an ideal milieu for Columbus, who could be
cushioned by the friendship of his fellow-countrymen, while in the service of foreign monarchs. This was to combine the best of two worlds.

What turned the Genoese dispersal around the Mediterranean into a
network, if not an empire, was not consciously imperial policy by the
Genoese State, but a sense—sometimes a muted sense—of national
solidarity, supplemented and often exceeded by family ties. In differing
degrees, this was characteristic of Mediterranean trading communities
generally. The outstanding example is formed by the Jews, who had no
state of their own but passed with ease from port to port or market to

market, among their co-religionists, and made their investments on the recommendations of brothers and cousins. Even in Venice, where commercial law was highly sophisticated by the thirteenth century and where people unrelated and even unknown to each other could form a mutual society or take part together in a share venture, most successful trading enterprises had a family basis. To be Genoese, however, was to belong to a community with distinctive features, and distinct advantages.

Genoese versatility did not exclude nostalgia for the metropolis. If the traders were successful in adapting to every economic environment and every political climate, the obverse of their ambivalence was an abiding sense of being Genoese and an enduring capacity to exploit Genoese connections. The street-names of fourteenth-century Kaffa—Genoa's sovereign Black Sea colony—recalled those of home. The poet known as the Anonymous of Genoa linked the adaptability of his fellow countrymen with their ability to replicate the 'tone' of their native city:

> So many are the Genoese
> And so extended everywhere,
> They go to any place they please
> And re-create their city there.[27]

This nostalgia may have been the basis of the expatriates' national fellow-feeling, which was to play a vital part in Columbus's career. Genoese abroad were naturally welcoming to one of their own. Columbus was, perhaps, the best-known beneficiary of this habit, first saved by Genoese in Lisbon, when he moved to that city probably in 1476 or 1477, then 'made' by Genoese of Seville, who deployed influence at the Castilian court on his behalf and raised money for his enterprises. Commercial considerations—it must be said—could outweigh the obligations of common provenance: a case in point is the cut-throat competition of the Genoese Centurione and Lomellini families, based respectively in Castile and Portugal, for a stake in the gold trade from the 1440s.[28] Only links of consanguinity or affinity were thick enough to provide an irrefrangible bond.

The same surnames crop up repeatedly all over the Genoese world, from the Black Sea in the thirteenth century to the Caribbean in the sixteenth. The Cattaneo, for instance, who were among the first great families to sprout a branch in Kaffa, were among the earliest Italian traders established in Mytilene; their relatives in Seville became collaborators of Colombus and they were also the first Genoese firm to open a branch in Santo Domingo. Transferable share-companies, like the

Maona (which possessed the monopoly of the exploitation of Chios) were rarer than the family firms, and even the Maona adopted for its members a common surname and some of the characteristics of a family. Every late medieval Genoese business which has been studied in detail has turned out to be in some sense a family business.[29] Thus, for Columbus, who was not a member of a merchant-clan, service with a merchant-house was an important, but limited, source of opportunity. Employment in the Centurione firm in the late 1470s, for instance, gave him independence from his own family and a start in the Atlantic trading-world; but it was unlikely to lead to the sort of wealth or fame to which he aspired, and though he was grateful to the Centurione for the start they gave him, which he remembered in his will, he got out when he could. The connection, however, proved of enduring value as the Centurione acted as bankers for the financing of his third trans-atlantic voyage in 1498, and continued to handle banking business for Columbus's heirs.[30]

The ambidextrous talents of Genoese traders made them adaptable not only to a variety of environments but also to a diversity of types of commerce. In the twelfth and thirteenth centuries the most attractive trades had taken them to the eastern Mediterranean, in search of spices. In the course of the fourteenth century, however, the Genoese switched the greater part of their effort (enormously the greater part in bulk terms and perhaps a little over 50 per cent in value terms) to the local products of the north-east Mediterranean basin, bulkier of carriage but depend-able of supply: above all, the mastic of Chios, the alum of Phocaea, Danubian and other northern forest products, the grain of Cyprus or the Danube or Black Sea basins and the slaves of the Black Sea. Spices, properly understood, tended to be channelled through Beirut and Alexandria, where the Venetians were supreme. The Genoese galleys went out of commission and were replaced—almost entirely by the end of the fourteenth century—by bulk-carrying round ships. At about the same time Chinese silks, a valuable perquisite of the trade of Genoese Romania in the early fourteenth century, became scarce as a result of the rupture of the 'Mongol Road' to China.[31]

As if to remedy these deficiencies, until under Genoese auspices Sicily and the Algarve began to yield commercial amounts of superior sugar and silk, Genoa found a silk- and sugar-land near home, at the western end of the Mediterranean, in the Moorish Kingdom of Granada. Though a Levantine rather than a fully Oriental spice, sugar was classified, with pepper, cinnamon, nutmeg, mace, and cloves, as an exotic condiment. Saffron and dried and preserved fruits were other Granadine products in much the same category. Granada's sugar industry had its own port at Almería, where most Genoese traders in the kingdom had representatives, but the main entrepôt was Málaga, an excellent harbour on the sea-route from the Mediterranean to the Atlantic, with access to a Granadine hinterland which served as a garden of exotic supplies.

Moreover, as well as being, as it were, a displaced Oriental land, Granada enjoyed privileged access to the Islamic Maghrib and therefore to Saharan gold—always the magnet and motor of European interest in Africa and the African Atlantic in the late Middle Ages. In the fifteenth century Málaga normally occupied third or fourth place among the Iberian ports which made direct shipments of Maghribi gold to Genoa. Seville, Cadiz, and Valencia were the other centres. But these statistics may obscure the primordial importance of the Kingdom of Granada in the gold trade. The gold, once seaborne, travelled by complex routes. The Genoese seem to have found it convenient to buy their gold in Castile and Valencia, where the silver price was relatively low. And while much of that gold, especially in Valencia, was the result of direct trade with Barbary, one of Castile's major sources was Granadine tribute, which must have passed virtually under the eyes of the Genoese of Málaga on its way to the hands of their brothers, cousins, partners, and bosses in Cadiz and Seville.[32] Throughout the period of Columbus's suit for patronage at the Castilian court, the monarchs of Castile were engaged in a war of conquest of Granada; Málaga fell to them a year after Columbus's first appearance before them; the last stronghold, Granada itself, almost on the day they decided to grant Columbus a commission. Against this background, the prominence given by Columbus to references to the gold trade in his correspondence with the monarchs is readily intelligible.[33]

All the industries served by Genoese trade implied geographical specialization, which in turn implied long-range commerce. The textile industry depended on a concentration of wools and dyestuffs in industrial centres; 'food-processing' on the meeting of fresh foodstuffs with salt. The gold industry of Genoa itself—turning raw gold into coins, leaf, and thread—relied on supplies of African gold to Italian technicians; shipbuilding demanded a similar marriage of raw materials with technical expertise, and a matching of wood, iron, sailcloth, and pitch. Part of the stimulus, therefore, for Genoese penetration of the Atlantic derived from the commercial needs and opportunities generated by

Genoa's presence in the eastern Mediterranean. And when Genoese colonial activity began in earnest in the Atlantic archipelagos—especially those of Madeira and the Canaries—in the fifteenth century, the eastern Mediterranean supplied vital economic models and new commodities, which would transform the islands' ecologies and form the basis of the early Atlantic economies.

The most powerful of these commodities was sugar. Uniquely among the exotic condiments favoured by palates in Latin Christendom, sugar could be grown in the Mediterranean. The first Genoese-owned sugar estates on a commercially influential scale seem to have been in Sicily, from where in the fifteenth century the crop was taken first to the Algarve, then to the Atlantic islands; here—in Madeira, the western Canaries, the Cape Verde Islands and those of the Guinea Gulf—it became the basis of the islands' economy by the end of the century.[34] By the time sugar completed the Atlantic crossing, planted by Columbus in Hispaniola, the effective model—it is usually supposed—was no longer the eastern Mediterranean but the Canaries. Yet it is worth remembering that Columbus's early career spanned the whole of this Genoese trading world, from Chios in the east to the Atlantic archipelagos in the west, and that he carried Mediterranean images in his mind. He claimed, for instance, that Hispaniola produced mastic: he must have been thinking of the island as another potential Chios where, he remembered, the trade was worth fifty thousand ducats a year.[35]

The main nurseries of Genoa's Atlantic experience were in the western Mediterranean, and especially in the Castilian empire of Andalusia, with its emporia of African gold and its deep-water harbours for Atlantic navigation. The nature of Genoese Atlantic colonization would follow the mercantile, small-scale, family-centred, ambivalent and 'stateless' tradition which typified the Genoese experience generally and monopolized that of the western Mediterranean. Just as in the fourteenth century the main theatre of Genoese commerce moved from Egypt and the Levant northwards to the Danubian provinces of the Byzantine and Ottoman empires and to the Black Sea, retreating from Venetian supremacy, similarly in the fifteenth a gradual displacement westwards was impelled by the rise of the Ottomans, who proved to be rapacious conquerors and unreliable partners in trade. By the end of the century, Chios was the only surviving Genoese sovereign possession in the east: it had become an entrepôt for the distribution of Atlantic sugar. Some adventurous spirits among the Genoese of the east had been tempted beyond the reach of the Turks into the Persian empire, India, and even Abyssinia. But there they operated without contact with their motherland. The main thrust of collective Genoese action reverted to Genoa's home waters and nearby bases, and therefore to the Atlantic. Genoa stood in similar geographical relation to the Atlantic, as Venice to the Orient. The Genoese, it seemed, had penetrated everywhere: there were places named after Genoese adventurers on the Sea of Azov and in the Canaries. But the Atlantic was their proper sphere.

Yet when the opportunity to exploit the ocean arose, Genoa lacked the resources, and especially the manpower, to make full use of it. Genoese expansion, despite its extraordinary powers of extension, was not infinitely elastic. Partly because of the drain of colonial enterprise, or simply because there was no more room to build inside the notoriously overcrowded city, Genoa's growth seems to have been checked. Working from the physical dimensions of the city and a reckoning of the number of buildings, Jacques Heers has calculated a mid-fifteenth-century population of over 100,000; his estimates of density, of heads per hearth and of hearths per house seem, however, to have been generous. Census figures of the sixteenth century suggest a total of only about a half of Heers' value—comparable, therefore, with Valencia or Barcelona rather than Venice or Seville.[36] The city Columbus left, untouched by the Renaissance, unenlarged by growth, relatively unadorned by expatriate wealth, was no longer the 'mistress of the seas' hailed at the height of her dynamism. In the Atlantic trade of the sixteenth century, Genoese no longer figured as pioneers or even, to any great extent, as participants. They were limited to a vicarious role, with Castilians as their main surrogates. Columbus was virtually the last such pioneer and his Genoese backers were representative of a new breed who preferred the bloated purse to the billowing sail. Their advantages and limitations—a talent for vicarious expansion, a severely constricted home base, a tradition of commercial conquest and of non-sovereign settlements—help to explain why Genoa made a vital contribution to the exploration and colonization of the Atlantic without establishing a sovereign Atlantic empire.

For Genoa's future in Atlantic trades, the most important of her merchant settlements were those in Castile and, particularly, Andalusia. For technical and geographical reasons, Cadiz and Seville and their regions were Genoese merchants' most substantial bases in Spain. The technical reasons were a matter of ships and cargoes. As early as 1216, James of Vitry had praised Genoa's big round ships, which could sail in winter and 'keep food and water fresh'—that is, exhaust shipboard supplies at a slower rate than the relatively heavily manned galleys. Thus the Genoese had large sailing vessels as well as galleys available to make the Atlantic run from the earliest days in the late thirteenth century. Commercial galleys disappeared from Genoa in the next hundred years.[37] It is probably safe to assume that the Atlantic routes, where conditions least favoured the galley, were the first to be converted to round ships. The use of round vessels, however, was not solely or even chiefly for the convenience of the navigators: Venetians sailed regularly to England and Flanders in galleys, and when the Atlantic commerce of Florence started in the fifteenth century, it was carried exclusively by galleys which demonstrated their worthiness for the task and continued to do so until the time of the Armada. The Genoese preference for more economical shipping was a result of their reliance on relatively low-value bulk commerce. The consequence was that a short run, direct from the mouth of the Mediterranean to the English Channel, was possible and even vital, since, by exploiting the seagoing properties of round ships to reduce the number of calls at ports en route and shorten the length of the voyage, merchants could ensure an improved return. Nor was there much point in hawking small quantities of the goods the Genoese carried; they were better held over for the big markets of the north, where the holds could be restocked with wool and cloth. Finally, the Genoese ships required spacious deep-water ports like those of Cadiz and the mouth of the Guadalquivir. It became normal for Genoese northbound ships to bypass Portugal, the Cantabrian Sea, and Atlantic France altogether.

Andalusia thus became a 'frontier' land of Genoa as well as of Castile. The colonization, which had begun before the Castilian conquest, grew enormously in the fourteenth century, once Genoa's northern commerce was established, and again in the fifteenth when opportunities in the east diminished. An idea of the scale and character of the colonies can be gained from a look at the region of Cadiz and Jeréz in the fifteenth century.[38] First, the growing pace of Genoese settlement and its changing, increasingly mercantile character are apparent. The first Genoese to settle in Jeréz, in the thirteenth century, for instance, were Benedetto Zaccaria, the celebrated naval commander, and Gasparo di Spinola, a retired ambassador. Zaccaria seems to have forsaken commerce during his residence in Castile and it was common for early Genoese colonists, especially in centres like Jeréz and Cordoba, in the hinterlands of the great ports, to marry into the local aristocracy and become *rentiers* rather than traders. In Seville and Cadiz the process was reversed and Genoese influence helped to convert the aristocracy to commerce in more than two centuries of increasing infiltration and intermarriage. By the time of the *nouvelle vague* of Genoese immigrants of the late fifteenth century, new settlement was exclusively mercantile and artisan. By the sixteenth century, three-quarters of the nobility of Seville had Genoese surnames; and their homespun philosopher, Jacopo Adorno, was there to justify for them the compatibility of commerce and nobility. As the pace of immigration grew, so did the amount of fixed settlement. Most Genoese newcomers of the fifteenth century tended to become 'citizens' (*vecinos*) as well as 'transients' (*estantes*) though the latter were preponderant: this presaged the relative immobility of the Genoese community of Castile in the sixteenth century, when they took part in Castilian empire-building vicariously, from their fixed centres, by way of banking and investment, rather than continuing to settle new frontiers.[39]

At the same time, these increasingly settled communities were in the fifteenth century points of support for further Genoese colonization westward in Portugal, in Africa, and above all in the Atlantic islands. Family homes established in Andalusia were ports of call for relatives from Genoa en route to or from the west. After taking up service with Portugal, for instance, in the exploration of the west coast of Africa, Antonio di Usodimare was sheltered by his brother Francesco in Cadiz in 1462. The Franchi di Luzzardo family sent sons on to Tenerife and Barbary, the Ascanio to Gran Canaria, the Nigro to Portugal and Madeira. Nothing better illustrates the elastic properties of the Genoese family as an instrument of colonization than this ability to maintain mobility while cultivating roots.

Finally, the Genoese of Andalusia demonstrated a typical Janus-like versatility. They were able to camouflage themselves in local society by intermarriage, formal naturalization, bilingualism, service to the community and the Crown, and even modification of the orthography of their names; and at the same time they could conserve in big centres like Seville and Cadiz 'another Genoa' of their own. Apart from their exogamous habits, their record in office-holding is the best indicator of their success in local acceptance worlds. Columbus's financial backers, for instance, Francesco da Rivarolo and Francesco Pinelli, were councillors of Seville and close confidants of the Crown; Francesco Adorno sat on the town council of Jeréz and Gianbattista di Ascanio and Christoforo Maruffo on that of Cadiz. At a humbler level, Agostino

Asilio was treasurer of his parish in Puerto de Santa María. Yet these positions of confidence in Castilian society were generally won without sacrifice of Genoese identity, especially in Seville and Cadiz where the Genoese colonies were defined by their ancient privileges and their proper consulates and quays. It was common to have a home in Genoa—Rivarolo's was cited as evidence of the invalidity of his naturalization—perhaps as a bolt-hole. Even fabled Genoese thrift served this sort of versatility: notarial records of Cadiz show them spending sparingly, outside trade, save on jewels, rugs, and small, movable luxuries. By the 1480s there was a particular Genoese confraternity in Cadiz, with its own chapel in the cathedral, and there may have been similar organizations elsewhere. The confraternity of the Name of Jesus in Jeréz, for instance, had been founded by Genoese tailors. In Seville, the Genoese 'nation' continued into the sixteenth century the custom of making collective addresses to the Crown. The effects of ambivalence were most vividly expressed by Columbus himself: 'Sirs', he wrote to the directors of the Genoese State Bank of San Giorgio—at a time when he was, admittedly, somewhat disillusioned with Castile—'although my body wanders here, my heart is continually in Genoa.' They may also have enhanced the sensibilities shown by the Castilian verses of one of the leading poets of fifteenth-century Seville, known as Francisco Imperial, and always described as 'native of Genoa, dweller in the most noble city of Seville'. He loved his adoptive city, 'choicest in the kingdom', praised the beauty of its women and the justice of its kings, exhorted and admonished it, in an explicitly Dantesque vision, to purge itself of heresy and vice; but he never forgot his native city either, and was mindful to the end of Ianos of Troy.[40]

The Genoese contexts between which Columbus moved—those of Genoa, where he began, and of Andalusia, where he ended up—are therefore vital to an understanding of his career. They help to explain both how and why he moved to the shores of the Atlantic, and what happened to him when he got there. Columbus's experiences as a navigator across the span of the Genoese world of his day, and the westward shift of his centre of operations, are documented in the record of five voyages, four of which are mentioned only in fragments of a letter written by Columbus in January 1495, long after the events referred to, apparently with the specific aim of convincing his correspondents—the Spanish monarchs—of the range of his practical navigational expertise. Still, all the claims made in the letter are inherently plausible. The earliest voyage—if it happened at all—must have taken place by early 1472. According to Columbus's account, it started in Marseilles, commissioned by the Angevin pretender to the Neapolitan throne, with the aim of capturing an Aragonese ship in Tunis harbour. 'The men that were with me mutinied', Columbus claimed, 'and determined to return to Marseilles', whereupon

realizing that I could not change their minds by force but only by means of some guile, I agreed to their demands and, after altering the way the compass lay, I made sail when it was getting dark. And the next day, when the sun rose, we were beyond Cape Carthage, while they were all sure that we were heading for Marseilles.[41]

The episode is unlikely to have been a pure invention, since partisanship of the Angevins cannot have been calculated to endear Columbus to Ferdinand and Isabella who had inherited the Aragonese claims. The setting of the tale is the Gulf of the Lion and the Tyrrhenian Sea—Genoa's home waters, which he sailed so thoroughly in his youth that he remembered the sailing directions in detail until the last years of his life.[42] The story of the deception of the mutineers has the flavour of a copy-book *sententia*, designed to illustrate a maxim of moral philosophy, but is typical of the way Columbus liked to see himself. He told similar stories of how he duped the crew on his first transatlantic voyage by tampering with the log, and of how he intimidated the natives of Jamaica during his last voyage by predicting an eclipse.[43] Whether strictly true or not, the story should be seen as true to the man and as part of his claim to a sort of natural sagacity, a deck-wise craft, which made up for his lack of formal schooling.

The tale of the Tunis voyage represents a glimpse of the first decisive shift of Columbus's life: from the weaver's shop to shipboard. The second such shift, from the Mediterranean to the Atlantic, is exemplified by the remaining evidence of his early voyages. The date of the move can no longer be pinpointed with precision, but must have happened around the mid-1470s and not later than 1477. The early tradition of his providential escape from piracy and shipwreck on a commercial voyage from Genoa northwards is too romantic and dramatic to be accepted incautiously: the self-image Columbus presented to early writers, as a divinely elected protagonist of great deeds, is suspiciously well served by the story. But whether by divinely orchestrated drama or by some prosaic and forgotten means, by 1477 Columbus had certainly removed from Genoa to Lisbon where he began a long period of habitual or intermittent residence in Portugal and a lifetime of navigation in the Atlantic. Nor is a miracle necessary to explain what, in the circumstances of Columbus's life, was a perfectly logical move. In travelling between the Mediterranean and the Atlantic, in switching to an Atlantic base of operations, and in undertaking further Atlantic voyages he was reflecting the common current features of Genoese experience in commerce and colonization.

Columbus's transfer to an Atlantic milieu did more for him than merely acquaint him with the practical problems of navigation which he would have to confront in attempting an ocean crossing. He embraced, as well as an Atlantic destiny, a flesh-and-blood bride, probably in Lisbon (but possibly in Madeira or Porto Santo). The date remains undocumented, but the balance of probabilities favours 1478 or 1479. This marriage was the biggest single step Columbus had taken towards the social respectability he seems to have craved. By some standards, it was a modest-enough step. Dona Felipa was technically a noblewoman, descended, on her mother's side, from a family with a long record of service to the Crown, the daughter of a lord who possessed the defining characteristic of feudal nobility: jurisdiction over vassals. Her father, Bartolomeo Perestrello, had occupied in his lifetime one of the smallest, poorest, and remotest fiefs in the Portuguese monarchy, the island of Porto Santo. Still, for the Genoese weaver's son it was an enormous leap. Columbus acquired by marriage a taste of the very form of nobility to which he could aspire to rise by deeds: a seaborne fief, acquired by feats of oceanic derring-do. His wife's father had been a modest model of the ennobling effects of seaborne adventure, a humble embodiment of the prominent theme of medieval chivalric literature. If an early biographical tradition can be trusted, marriage also brought Columbus access to the papers of his bride's late father, which, according to tradition, stimulated Columbus's interest in the Portuguese record of Atlantic discovery.[44] Dona Felipa performed two further services for her husband: she provided him with his only legitimate son, Diego, on whom his dynastic ambitions became focused and in whom they eventually became fulfilled; and she died early, leaving Columbus free and, it seems, unencumbered by sentimental memories. His only affectionate reference to his wife occurs in a summary of his services to the Spanish monarchs, in which he speaks of having to 'leave wife and children' to come to their court.[45] But if this implies tenderness, rather than merely exemplifying Columbus's way with emotional rhetoric, it has to be acknowledged that the reference may be to a period after Dona Felipa's death and that the 'wife' in question was linked to Columbus by less formal bonds. His marriage provided him with one other potentially useful contact we know about: relations of his wife's—a sister- and brother-in-law called Violanta and Miguel Muliart—lived in Huelva, a stone's throw from Columbus's future point of departure for the New World. He visited them there in 1491, at a time when he was forging links with the seafaring community in nearby Palos. It would be tempting, but unsound, to infer some causal connection between the two events. It is more likely that the family link redounded to the benefit of the Muliart household, who borrowed money from their fortunate kinsman when Columbus became rich.[46]

The framework of his life in the late 1470s, during the period of his earliest Atlantic voyages, is fixed by a Genoese document of 1479 which records a voyage made by Columbus to Madeira the previous year to buy sugar as part of a deal set up by the Centurione firm.[47] The promoter Luigi Centurione and the intermediary in the deal, Paolo di Nigro, were remembered in the last codicil to Columbus's will many years later, together with various residents of Lisbon, including another member of the Centurione clan.[48] Employment as a sugar-buyer for Centurione family interests in the Atlantic islands would also provide a context for the trip between Lisbon and Porto Santo, presumably during the same period, which Columbus recalled in his letter of 1495,[49] as well as for visits to the Canaries and the Azores, which are not recorded in any document but which can safely be inferred from his obvious familiarity with both archipelagos, corroborated, in the case of the Azores, by the early biographers' claim that he gathered evidence of the navigability of the Atlantic in those islands.[50]

These three archipelagos—the Madeira group, the Canaries, and the Azores—were linked by wind-system and trade-routes to a wider circle of Atlantic navigation which stretched south to the Gulf of Guinea and north to England and beyond. In his reminiscences of 1495, and in hand-written notes in the margins of books he read,[51] Columbus recorded voyages to those extremes. He claimed that February 1477— the date can be treated as unreliable in such a long-deferred recollection—he sailed 'a hundred leagues beyond' Iceland, on a trip from Bristol, as the context implies; and in or after 1482, according to the same sources, he went south from Lisbon to the new Portuguese trading-post of São Jorge da Mina, near the mouth of the Volta, where the Portuguese gold trade with the inland mining centres was concentrated. The first of these routes was commonly sailed by Mediterranean merchants as far as England, where Columbus might well have joined a Bristolian venture to Iceland. There is nothing inherently implausible in his claim, and his participation in such a voyage would also provide a context for the visit to Galway in Ireland which he mentioned in another marginal annotation. The authenticity of his journey to the Gold Coast is abundantly confirmed by the familiarity shown with equatorial Africa at various points in his surviving writings.[52]

Thus, by the mid-1480s Columbus had very nearly justified his later boast to have sailed 'every sea so far traversed'. In particular, he had followed the course of Genoese commercial expansion from the Mediterranean to the Atlantic—almost spanning the Genoese world from Chios to the Canaries—and he had penetrated to the remotest corners of the Atlantic as it was known in his day.

The Atlantic—the 'Ocean Sea'—was an enticing world of opportunity for Columbus's contemporaries. The excited speculation aroused by the ill-defined expanse of unexplored ocean can be observed in maps of the time, which show what a stimulus to the imagination Atlantic exploration was, and how consciousness of a potentially exploitable Atlantic 'space' grew in the century before Columbus's voyages. To the mythical islands commonly asssigned imaginary positions in fourteenth-century maps—those of St Brendan, St Ursula, and Brasil—a Venetian chart of 1424 added large and alluring islands, including 'Antillia', identified with the island of 'Seven Cities' to which, in a legend not unlike that of St Ursula, Portuguese refugees from the Moors were held to have repaired in the eighth century. These islands became common in subsequent cartographical tradition and inspired voyages in search of them: Columbus sailed with them in mind. His island-discoveries were named the 'Antilles' collectively, and one archipelago he named after St Ursula's 'Virgins'. As late as 1514, Portuguese official sailing directions gave courses to islands 'not yet discovered' and one of the most amusing forgeries of the sixteenth century is a spurious Spanish 'chronicle' of the conquest of St Brendan's Isle.[53]

Although attempts have been made to relate these mental divagations to real finds, usually in connection with theories of pre-Columbian discoveries of America, the only possible new discovery of the early fifteenth century which might fit is that of the Sargasso Sea. But once one appreciates the genuine excitement aroused in the fifteenth century by the unlimited possibilities of the Atlantic, the fertility of speculation seems adequately explained. Fresh discoveries were a direct stimulus:

the Majorcan cartographers who first placed the Azores in their roughly correct position in maps of the 1430s also introduced new speculative islands into the tradition. Andrea Bianco of Genoa was interested in the latest verifiable novelties, as his map of 1448 shows, but in his world map of 1436 he scattered a handful of imaginary islands about the Ocean, and even in the 1448 chart he included some traditional isles, with an assurance that an 'authentic island' lay 1,500 miles out in the equatorial Atlantic.[54]

The voyages inspired by such speculation sometimes reported genuine new discoveries in their turn, which fuelled the process. Columbus grew up in a period when the Atlantic was being strewn with new landmarks and defined by newly revealed boundaries. In 1452—roughly at the time of his probable birth—the two remotest islands of the Azores were discovered. Between the mid-1450s and mid-1460s the Cape Verde archipelago was explored. In the 1470s the islands of the Gulf of Guinea were added. And during the 1480s, when Columbus was seeking to make an Atlantic voyage of discovery of his own, Diogo Cão and Bartolomeu Dias traced the west African coast to its southernmost limit. In 1480 and 1487, certainly, and perhaps regularly in the 1490s, expeditions sailed from Bristol in search of new islands: a big increase in the importation of North Atlantic products to Bristol in the 1480s shows the increased trade with Iceland that such journeys produced or reflected, but these were consciously exploratory voyages, intended to 'search and fynde'.[55] The Bristolians called their objective 'Brasil'. For the Portuguese and the Flemings of the Azores, 'Antillia' was a comparable catch-all name for potential new discoveries. At least eight Portuguese commissions for the discovery of new Atlantic islands survive from the years 1462–87. Some specifically refer to the evidence of sea-charts. The most general terms are those of Fernão Teles's grant (1474) of 'the Seven Cities or whatever islands he shall find'.[56] Despite the minimal results, the Bristolian and Azorean voyages continued unabashed. The Atlantic was becoming a compulsion, a vacuum irresistibly abhorred.

The pace of change in the received picture of the world seems to have made Andrea Bianco, for one, feel that ancient geographical certainties had to be discarded. The same point was made a few years later by the acknowledged master of the Venetian cartographic school, Fra Mauro, who confessed in a note on his world map—the fullest then devised—that his delineation must be imperfect, since the extent of the world was unknown. The cosmographers of the fifteenth century remind one of the prisoners-of-war in one of Marcel Ainé's stories, who, unable to see their cell walls in the light of a tallow candle, could imagine themselves free. The fertile ignorance of the late Middle Ages imparted a similar sense of boundlessness. Columbus, despite his slavish respect for selective written authorities, always exhibited a childlike delight when he was able, from experience, to challenge received wisdom. He was indebted to the confidence of some of the theoretical geographers of his time, who perceived the liberating effects of progress in exploration. He noted, for instance, Pope Pius II's arguments in favour of the navigability of all oceans and of the accessibility of all lands, and claimed that the Portuguese exploration down the coast of Africa had exploded ancient notions of the impenetrability of the torrid zone of the world.

In his ambitions and in his explorer's vocation, as in most other respects, Columbus was a representative figure of his day. The image of the lonely man of destiny, struggling against prevailing orthodoxy to realize a dream that was ahead of its time, derives from his own self-image as a friendless outsider, derided by a scientific and social establishment that was reluctant to accept him. To explain his unique achievement—the discovery of America—it is not necessary to suppose that he started with a unique plan, or a unique vision, or a unique pattern of previous experience. Columbus's design for a new Atlantic voyage clearly belongs in the context of an age of vigorous speculation about the secrets of the Ocean Sea. Almost every element in the thinking that underlay his enterprise was part of the common currency of geographical debate in his day.

'The Secrets of this World'

THE FORMATION OF PLANS
AND TASTES
c.1480–1492

FERNANDO COLÓN, Columbus's younger and illegitimate son, inherited a little of his father's adventurous spirit and a great deal of his bookish tastes. His enormous library—reputedly of over fifteen thousand volumes, some four thousand of which were minutely catalogued, with details of their contents—was one of the outstanding scientific collections of its day, especially in navigation and mathematics. But on Fernando's death in 1539 it passed into the hands of his wastrel nephew, Don Luis, and its dissipation began. In 1551 the cathedral chapter of Seville secured the reversion of the library, in accordance with a clause of Fernando's will, because of the heir's neglect. The new custodians took little more care of it; but among the fragments which survive to this day in a chamber of the cathedral dependencies, off the Court of the Oranges, are a few books which belonged to Columbus, four of which are scrawled with his marginal annotations. These remnants of his thoughts lie close to the reputed resting-place of the remains of his body, in the transept of the same cathedral. They form a maddeningly oblique but irresistibly inviting route of access into the process of Columbus's self-education and into the formation of his project for an Atlantic crossing. The information they yield can be supplemented from references to Columbus's reading-matter both in his own writings and in the accounts of contemporaries, but it remains intriguingly fragmentary and ensnaringly hard to interpret. Though quarried for evidence of the sources of Columbus's cosmography, the annotations actually reveal more—as we shall see—about his values and tastes.

The danger of these sources is that they encourage us to see

Columbus as an intellectual—which he was, in a small way, but not to the exclusion of his vocation as a man of action—and to see the formation of his projects as an academic exercise. Columbus was a much more mercurial creature than is commonly credited, but one belief he held with absolute consistency was in empirical epistemology. One learned above all, he claimed, from experience, or as he once put it, quoting a proverb, 'as one goes, one's knowledge grows'.[1] He felt that his own practical seafaring and cosmographical learning were inter-dependent. The 'very occupation' of a mariner inclined men 'to seek to learn the secrets of this world'[2] and book-learning could be practically applied. Columbus claimed, for instance, that the Caribbean would be impassable to navigators who lacked the astronomer's esoteric art.[3] I think it can be shown—I hope it will be shown in the course of this book—that the impact of the experience of the New World affected Columbus's ideas, and even modified his geographical theories, after 1492; and it is extremely well attested that before that date Columbus supported his Atlantic project both with evidence garnered from his personal knowledge of the ocean and with references to written authorities. Examination of what might be called the literary sources of Columbus's plans should not be made without remembering his long and extensive experience of the ocean, which alone might have been sufficient to inspire in him a desire to cross it.

Most studies of Columbus follow his sixteenth-century eulogists who have assumed that the complex justifications of his enterprise which he wrote from 1498 onwards were already elaborated before his first crossing, with all the classical, apocryphal, patristic, and medieval sources docketed in their place. He certainly began to acquire a literate culture well before 1492: his temporary change of vocation—at an undetermined date, probably fairly early in the second half of the 1480s—from mariner to bookseller evokes the alchemy that transmuted Columbus, the merchant's agent, into Columbus, the learned geo-grapher.[4] Whether or not his reading contributed to the formation of his plan, it certainly contributed to its presentation: according to Columbus's own later memories, he expounded his ideas before the Spanish monarchs with the aid of maps and books.[5]

The assumption, however, that the learning Columbus called on from 1498 was already fully mastered by 1492 is extremely rash. The theory, for instance, ascribed by Ptolemy to Marinus of Tyre, that the Eurasian landmass extends for over 225 degrees on the surface of the world, has normally been supposed to have been part of Columbus's early intellectual armoury. This would be convenient, if it were so, because it would help to explain why Columbus might have thought the Atlantic was navigably small, but he never mentioned Marinus until after he had conducted a series of relevant experiments—beginning in western Hispaniola in 1494 —in an attempt to calculate the breadth of the ocean in terms of degrees. ˙ ˙e possessed a fairly heavily annotated copy of the *Historia Naturalis* of Pliny, to which he made reference in connection with the identification of the mastic plant in 1492; but a note referring to Hispaniola in the margin of his copy shows that he continued to read or reread the work after that date, and again it was not until 1498 that Pliny was used by Columbus in the context of the exposition of cosmographic theory. Again, the occasion was prompted by the need to consider empirical evidence, which, Columbus thought, appeared to call in question Pliny's views on the sphericity of the earth.[7] From surviving examples, we know that Columbus continued to acquire books at least until 1496, the year of publication of his copies of the *Philosophia Naturalis* of Albertus Magnus and the *Almanach Perpetuum* of Abraham Zacuto; in the same year he sent to England to obtain a copy of *Marco Polo*.[8] Much of the reading to which references can be traced in his writings could have been done during his periods of enforced leisure in Spain in 1496–8 and 1500–2: the burden of erudition is most heavily borne in writings from, during, or just after those periods. At almost every discernible stage of his career from the 1480s onwards, practical experience and book-learning seem to have reinforced one another, with neither monopolizing the processes of his intellectual formation or mental development. It would be equally convincing to argue that he tended to turn to written authorities for confirmation of, or glosses on, ideas prompted by experience as to claim that his achievements in Atlantic navigation were made in the course of applying academic theories. The truth probably lies in a mixture of both points of view.

Practical experience takes a long time to acquire; nor, for a self-educated man, does erudition come easily or quickly. Of any body of ideas, time tends to change the shoreline, by erosion in some areas and by accumulation in others. It is important not to be deluded by Columbus's reputations for obstinacy and for unshakeable self-faith into supposing that he was incapable of changing his mind. He could refine and even reverse ideas; nor should it be forgotten that for six or seven years at the Castilian court he was virtually in the position of a professional lobbyist, responding to the need to modify the presentation of his views as he addressed different potential patrons and intermediaries. Columbus himself was insensible to his own mutability; selective memory and tendentious presentation made his intellectual formation, in all his own accounts, a *coup de foudre*. He adopted a model from hagiographical literature, in which he was presented with a previously obnubilated truth by 'God's manifest hand',[9] and subsequently never ceased to advocate it. Yet—to substitute a classical image for a hagiographical one—such wisdom as Columbus's does not commonly spring fully armed from the head. His basic project for a crossing of the Atlantic, and the geographical ideas which sustained it, are more likely to have emerged slowly and matured gradually. A transatlantic project could be presented in various ways—for instance, as we shall see, as a quest for new islands, a drive to Asia, a hunt for a new continent—and could be associated with a variety of possible objectives. To it could be added a remoter 'grand design' to take Islam in the rear and reconquer Jerusalem, such as Columbus first advocated before 1492, and returned to at intervals, developing it, as time went by, in increasingly eschatological terms, with increasingly millenarian constructions.[10] Without appreciating the potential for change and development in Columbus's ideas, it is impossible to give a useful account of them.

Nor can they be understood by accepting his image of himself as a singular figure, uniquely gifted with divinely derived insights. His geographical ideas were neither unchanging nor unrepresentative of their time. Before examining in detail the intellectual influences to which he was exposed, it will be helpful to sketch in the context of freedom of conjecture about the Atlantic, shared by mapmakers, cosmographers, and—presumably—explorers in fifteenth-century Latin Christendom. Against this background, Columbus's scheme for crossing the ocean seems comfortingly intelligible, even predictable.

It was a period in which Atlantic space exerted a strong pull on imaginations in Latin Christendom. Mapmakers scattered their representations of the ocean with speculative lands and, from 1424, left blank spaces to be filled in with new discoveries. As interest in the space grew, so did consciousness of its exploitability. The first enduring colonies were established in the Canary Islands in 1402, in the Azores in 1439. The pace of endeavour quickened in the second half of the century. The island of Gomera was conquered and Flores, Corvo, the Cape Verde Islands, and the isles of the Gulf of Guinea were explored in the generation after 1450. Cartography, lagging behind discovery, did not incorporate all of these—or, indeed, delineate previously known islands with perfect accuracy—until the 1480s. Yet, while slow to reflect discovery, maps were quick to encourage it: voyages launched from Bristol in the last years of the century, to find the cartographers' isle of Brasil, demonstrate this, as do those of the Portuguese of the Azores to find the similar invention known as 'Antillia', or those of Columbus himself, which were partly guided by a speculative map. The conquerors of the Canaries in 1402 were lured into the ocean by a cartographer's notion of a 'River of Gold'. Even in world maps—which gave their compilers a welcome chance to speculate about the Orient—the greatest concentration of novelties, after those drawn from Ptolemy, lay in the Atlantic. The extent of speculation about the ocean is one of the most remarkable features of the cartography of the time. Until the work of Columbus defined its limits, the possibilities of the Atlantic were alluringly boundless.[11]

Apart from the common belief that it concealed more undiscovered lands, two speculative theories about the Atlantic, current in his day, had a direct bearing on Columbus's own project, the theory of the existence of the Antipodes, and the theory of a narrow Atlantic. Although each had a long pedigree, there was an avant-garde feel to both of them. They were both concerned with the same basic problem—the size of the globe—and both theories were developed in response to it. The girth of the world had been consistently underestimated since antiquity, though the best available computation, that of Eratosthenes of Alexandria, was accurate to within at least 5 per cent—perhaps to within 2 per cent if the most favourable values are assigned to the cosmographer's units of measurement. Eratosthenes had used a theoretically foolproof method, calculating by trigonometry the angle subtended at the centre of the earth by a measured line between two points on the same meridian. In practice, however, the method involved a generous margin of error: the distance between the chosen points was hard to measure accurately; and there was almost bound to be some difference, if only a slight one, between their true respective meridians.[12] Thus, while the Alexandrian's method commanded admiration, his results were open to doubt.

Even the common underestimates of the size of the world implied a huge unknown moiety, the *pars inferior*, concealed from scrutiny like the dark side of the moon. As the image of the *orbis terrarum*—the single, continuous landmass comprising all the known world—was firmly etched in the mind of every educated man, the common, received wisdom was that unrelieved ocean occupied the unknown portion. The daring thought that there might be a second landmass in the midst of the ocean 'opposite' the familiar world appealed to the Renaissance taste for symmetry and, more generally, to the medieval preference for an ordered and 'concordant' creation; but it breached two dearly held shibboleths: that all men were descended from Adam and that the apostles had preached 'throughout the world'.[13] Belief in the Antipodes in the late Middle Ages can fairly be compared with belief in the existence of inhabited worlds in outer space today: both types of world were fervently imagined and sceptically dismissed.

Yet the possibility of the existence of the Antipodes was increasingly widely canvassed. In the early fifteenth century Pierre d'Ailly, the reforming Cardinal of Touraine, referred to it in his *Imago Mundi*, one of the most influential cosmographical works of the period, and in two treatises written a few years later under the influence of Ptolemy. In his *Historia Rerum* of the mid-fifteenth century, Æneas Sylvius Piccolomini (the future Pope Pius II) gave the theory his implicit approval, only to grant it a pious dismissal with a reminder that a Christian 'ought to prefer' the traditional view: clearly, that sort of disclaimer was not to be taken seriously. Columbus's copies of both works survive, bearing the signs of thorough perusal. By his time, the existence of the Antipodes was widely debated and, in some circles, especially in Italy and among humanists, their discovery was seriously anticipated. Under the name of 'Hesperides' they appeared on some maps.[14]

Speculation about the Antipodes may have been stimulated by the reception of Strabo's *Geography* in the West. The text arrived in Italy in 1423 and some of Strabo's ideas circulated widely from the time of the Council of Florence in 1439—a great occasion for the exchange of cosmographical news as well as ecclesiological debate. A complete translation of the *Geography* by Guarino of Verona was available from 1458 and in print by 1469. Its peculiar importance was that this text placed the supposed unknown continent roughly where Columbus or one of the other Atlantic navigators of the fifteenth century might have been expected to find it: 'It may be that in this same temperate zone there are actually two inhabited worlds or even more, and in particular in the proximity of the parallel through Athens that is drawn across the Atlantic Sea.' In the context of Strabo's thought generally it seems that this observation may have been intended ironically; but irony is notoriously difficult to detect in texts from an unfamiliar culture and Columbus's contemporaries took the passage literally. It was striking in view of Strabo's general defence of a Homeric world picture and rebuttal of the cosmography of Eratosthenes: in particular, Strabo intended it as a challenge to Eratosthenes' view that 'if the immensity of the Atlantic Sea did not prevent it, we could sail from Iberia to India along the same parallel'. Columbus cannot be proved to have read Strabo, but the cartouche of a map attributed to his brother Bartolomé and known to sixteenth-century writers cited the geographer as well as Ptolemy, Pliny, and Isidore.[15]

That Columbus himself considered the Antipodes as a possible destination for his own projected Atlantic exploration is suggested by the response of one of the committees which investigated his plans: 'St Augustine doubts it.' This looks like an allusion to St Augustine's doubts of the existence of the Antipodes. When Columbus returned from his first voyage, despite his energetic avowals that he had been to Asia most Italian commentators seem to have assumed that his discoveries were Antipodean in character: the ready acceptance of this theory in so many sources shows that it must have been current before Columbus's departure. Evidently, its attraction for humanists was particularly strong, perhaps because it appeared to draw support from authorities widely esteemed for their place in the classical tradition, such as the *De Nuptiis Philologiae et Mercurii* of Martianus Capella or the commentaries on Cicero of Macrobius, whose world picture seems to have been largely based on Eratosthenes and therefore to have belonged to a quite different school from Strabo. Macrobius—albeit more obliquely than Strabo—also suggested that an 'Antipodean' landmass might exist in the northern as well as the southern hemisphere.[16]

The second great theory was that a greatly extended world landmass filled the Atlantic space, leaving relatively less of the globe's surface to the intervening ocean. Ptolemy had mentioned dismissively the calculations of his fellow cosmographer, Marinus of Tyre, who had sought to stretch the limits of Asia eastwards beyond those acceptable to Ptolemy. Following this hint, Pierre d'Ailly speculated that the Antipodes might not be a separate continent, but contiguous with the known landmass. D'Ailly passed on a range of authorities collected by Roger Bacon (1214–92), perhaps with some distortion of the authors' original intentions, to suggest that most of the surface of the world is covered by land: a narrow Atlantic was a necessary inference drawn by some of d'Ailly's readers, including, explicitly, Columbus. In attributing to Aristotle the view that 'the sea is small between the western extremity of Spain and the eastern part of India', d'Ailly was being more faithful to Bacon than to Aristotle, whose text on the subject is ambiguous and obscure. But such authority, once appropriated or arrogated for a particular point of view, carried enormous weight. At the time of his third voyage Columbus cited it repeatedly in support of his claim to have reached, or closely approached, Asia.[17]

The theory of a narrow Atlantic was cultivated in the circle of the Florentine cosmographer, Paolo del Pozzo Toscanelli, whose views were expressed in a letter of June 1474 addressed, via a canon of Lisbon, to the Portuguese king, and in a subsequent recapitulation, of uncertain authenticity,[18] addressed to Columbus. Toscanelli estimated the distance from the Canaries to Cathay to be about 5,000 nautical miles: not quite navigable by the standards of the time, but the journey might, he thought, be broken at 'Antillia' (the mythical island of Portuguese tradition) or at Japan, which, on Marco Polo's authority, was thought to lie at a huge distance from China. As copies of the Toscanelli correspondence survive, in Columbus's hand, bound into the endpapers of one of his books, there can be no reasonable doubt of his awareness of these views. The date at which they became available to him is, however, a moot point. That he received them before 1492 can be regarded as probable, but not certain. At the very least, they show the sort of projects which were in the air before Columbus's departure, and the diversity of views about the nature of Atlantic space.[19] Toscanelli's picture of the Atlantic, or a version very similar to it, soon came to be shared by cosmographers in Nuremberg: it is, with only small modifications, the picture represented on a globe made in that city by Martin Behaim in 1492 (see Map 2); the following year Hieronymus Münzer wrote from there to the King of Portugal, urging the exploration of a western route to Asia.[20] By then, of course, apparently unknown to Münzer, the attempt had already been made.

Even Toscanelli's Atlantic was, for practical purposes, unnavigably broad. Columbus, however, proposed to shrink it conceptually by arguing that 'this world is small'. In his surviving writings, Columbus did not specifically address the size of the globe until very late. His first discussion of the problem was written in August 1498.[21] It is fair to suppose from the weight of circumstantial evidence, however, that from a much earlier date he shared or exceeded his contemporaries' tendency to underestimate. The materials on which his calculation was based were almost all drawn from the *Imago Mundi* of Pierre d'Ailly, which he had probably read for the first time by 1488—the date of the earliest securely datable marginal annotation which he or his brother made in the text.[22] When he came to expound his views in detail, he espoused an underestimate more grossly distorted than any ever recorded: 25 per cent below the true figure and at least 8 per cent below even the most adventurous estimate otherwise known to have been made in his day. His avowed basis for this calculation was obviously mistaken: he claimed, in an undated marginal note to his copy of Pierre d'Ailly's book, that his own comparisons of observed latitudes and recorded distances during a voyage to the Gulf of Guinea had convinced him 'that my measurements endorse the opinion of Alfraganus: that is, that to any one degree, 56⅔ miles correspond . . . Therefore we may say that the perimeter of the earth at the equator is 20,400 miles'. He went on to claim that his readings were confirmed by Portuguese experts, including the well-known cosmographer José Vizinho.[23]

This last claim is incompatible with other evidence about Vizinho's opinion and the rest of the note creates a highly misleading impression. 'The opinion of Alfraganus'—the tenth-century Arab cosmographer al-Farghani—was expressed in miles which were much bigger than those of authorities in the Greek and Latin worlds: Columbus, who got the information at second hand from Pierre d'Ailly, failed to take the elementary precaution of standardizing his units of measurement. And even had his figures been right, they could not have been verified in the way Columbus claimed—using 'a quadrant and other instruments'. The inaccuracies of this method of reading latitude at sea were never resolved in the fifteenth century; mariners' calculations of distance were extremely rough-and-ready—the estimates of Columbus's pilots on his first transatlantic voyage, for instance, varied by as much as 10 per cent; and in any case Columbus could not be sure that his route to Guinea was taking him along a great circle of the earth.[24]

It would be unsound, therefore, to suppose that Columbus had formed his mental picture of a small world as early as the date of his Guinea voyage, which can be traced to between 1482 and 1485, with some reason for preferring the last year.[25] His memories of that experience could have been affected by later events and the recollection modified accordingly. It would, for instance, be consistent with the evidence to imagine Columbus rereading Pierre d'Ailly attentively, in or about 1498 when he was preparing his detailed vindication of his claim to have discovered a short route to Asia; he might then have attributed to the time of the Guinea voyage the inception of views really espoused later. None the less, he certainly had access to, or had formulated for himself, some sort of 'small-world' theory or at least a narrow-Atlantic theory—perhaps Toscanelli's—by 1492. Otherwise, his repeated advocacy in writings that can be dated to that year of the proximity of Asia to Europe travelling westward would be inexplicable.[26]

While apparently contemplating destinations in Asia and the Antipodes, Columbus also seems to have kept a third possible objective in mind. At the very least, according to an early and privileged biographer, he was hoping to add to the growing list of newly discovered Atlantic islands. Much of the empirical evidence which he gathered concerning the potential of the remoter Atlantic related only to unidentifiable new lands. Although, for instance, the flat-faced castaways he claimed to have seen in Ireland must, he thought, have come directly 'from Cathay', the indications from assorted flotsam and driftwood washed up on Atlantic shores were indifferent: they could have come from any land lying westward. Even more emphatically, the reports he gathered of mariners' sightings of evanescent islands—probably mere cloudbanks—beyond the known Atlantic archipelagos nourished expectations of new discoveries to be made. His work as a cartographer made him thoroughly familiar with the fabulous Atlantic gazetteer of his fellow-professionals: he had a 'map of islands' with him on his first crossing of the ocean, and treated it with the respect due to a reliable source.[27]

Thus, during the period when he formulated his Atlantic design and sought patronage for it, in the 1480s and early 1490s, Columbus could have had three possible destinations in mind: Asia, the Antipodes, and as yet undiscovered islands. Historians and biographers have generally been anxious to pin him down to one or other of these, following the tradition inaugurated by Columbus himself who seems to have regarded consistency as evidence of divine election. The objective evidence, however, suggests that he considered all three destinations at different times, or sometimes simultaneously, and advocated them severally in addressing different audiences. The terms of the commission he at last obtained in Spain refer to 'islands and mainlands' as his objective— a phrase which covers all possibilities; 'St Augustine's doubts' point to a project for the discovery of the Antipodes; the early tradition that his proposals in Portugal lacked credibility in part because of 'his fantasies with his island of Cipangu', suggests, if reliable, that he envisaged an Asiatic destination in the earliest phase of his quest for patronage.[28] Once it is remembered that Columbus could have embraced exploration not for its own sake but as a means to personal social advancement, then the need to see him as consistent in his specific projects for voyages disappears. His 'certainty' vanishes. He was determined to make a voyage, but prepared to contemplate a variety of destinations.

Columbus's mental world-picture, his geographical ideas in general, took shape between the beginnings of his self-education in cosmography, probably in the 1480s, and the period of his systematic writings on the subject, from 1498. To be more exact, the start of the process can be tentatively traced to before 1484, when he is traditionally supposed to have made his first submission to the Portuguese Crown, and its completion or maturity to before 1495, when he already had a reputation for cosmographical learning among contemporaries.[29] In any event, if one discounts the quasi-hagiographical tradition that sees his ideas as fully formed before his first Atlantic crossing, and sets aside the equal but opposite case of detractors who suppose that Columbus must have received his world-picture, all at once, from some unknown originator, the process of his intellectual formation was a long one, spanning his career as a transatlantic navigator, and was fed by his experiences and observations as well as by his reading. Like religion, learning was something which grew on him as time went on; his reactions to what he read were warped by his own triumphs and afflictions. In consequence, his geographical ideas constantly evolved and could sometimes be dramatically revised. The adamantine Columbus we have inherited from tradition needs to be rebuilt in mercury and opal. He was temperamentally obstinate and obsessive, no doubt; but he could be so successively about different ideas.

It is probably fair, none the less, to see the 1480s as a crucial decade in the formation of Columbus's intellect, when he became a geographer as well as a practical navigator, and when he acquired enough learning to be able to support projects for exploratory voyages with arguments drawn from written authorities. For at least part of this period Columbus abandoned his merchant's lot and plied the trades of bookseller and mapmaker: the early testimony of Andrés Bernáldez, who knew him well, and Bartolomé de Las Casas, who had access to his papers, is supported—at least as far as the maps are concerned—by Columbus's own boasts that God had taught him the cartographer's art and that he had displayed maps to his patrons when supplicating for support.[30] If this evidence is reliable, Columbus had acquired a bookish side to his character by the period in question, which must, from the context of Bernáldez's and Las Casas's claims, have been early in the second half of the 1480s. From his inception in the trade he had privileged access to books. The tradition that his brother Bartolomeo (always called, in Castilian orthography, Bartolomé Colón) had joined him in Lisbon where he had learned to make maps may help to explain Columbus's change of trade and suggests that the beginnings of their joint venture in a literate craft might be dated even earlier; for Columbus's period of habitual residence in Lisbon cannot have extended beyond 1485.

In the brothers' handful of surviving books the annotations scrawled in the margins, in crabbed and almost indistinguishable hands, provide our only first-hand evidence about the formation of Columbus's intellect. In recent years they have been scrutinized by scholars with greedy efficiency.[31] The difficulties of dating books which may have been read and reread many times over a long period raise the danger of our assigning to an early period of Columbus's life beliefs formulated and concerns shaped only much later.

Some of the priorities they reveal are hard to fit into a convincing picture of Columbus's development. His interest in questions of hydrography, for instance, is obvious, but seems relevant only to his vocation as a cartographer, not an explorer. He was evidently obsessed by the legend of the Amazons, taking note of every reference he encountered; and indeed, twice during his explorations in the New World he thought he had encountered, or narrowly missed, such beings. But was a search for the Amazons part of his 'grand design'? Was he interested in them merely as a decorative motif for his putative maps? Did he see them as a source of flattering images to be applied to the Queen of Castile, who perceived herself as a *femina fortis*? The annotations also show his interest, almost certainly formed before 1492, in computations of the age of the world and, by implication, in calculations of the date of the millennium. Millenarianism became, from the late 1490s, one of Columbus's most prominent pet obsessions, and the conquest of Jerusalem, which he claimed to have proposed to the Spanish monarchs as a future project before his departure for the New World, was treated as an eschatological symbol in his later writings. But does this mean Columbus was already a millenarian fantasist, nourishing a chiliastic 'hidden agenda' before 1492? It is prudent to assume no more than that the early annotations are guides to trends in Columbus's thinking, which may have begun early and matured late.

Of the half-dozen books he read most thoroughly, and which can be surmised to have had some possible influence on him, at least four must have been in his hands before 1492. It was only in the late 1490s, however, that Columbus began (in surviving writings) to weave their information into what might be called a systematic cosmography; it would be rash to suppose that Columbus in, say, 1498 was merely recapitulating ideas consistently expounded since before 1492. None the less, the books in question remain an invaluable guide to the range of ideas available to Columbus during the formulation of his design. They were in all likelihood the bedrock of the written support for his case which he later recalled submitting during his campaign to secure patronage:

the writings of many trustworthy authorities were cited, who wrote works of history in which they said that there were great riches in those parts of the world. And likewise it was necessary to bring to bear on this the sayings and views of those who had described the geography of the world in writing. And at last your Highnesses resolved that it should be put into effect.[32]

Of these essential texts, none was more important, either for Columbus or generally for the geography of his time, than Ptolemy's *Geography*.[33] Rediscovered by Western scholars early in the fifteenth century, this second-century Alexandrian compendium reunited much classical learning and speculation. In Italy and Portugal, where Ptolemy had enjoyed his longest and widest currency, the authority of the *Geography* was thought superior to that of all other texts. At Ptolemy's remote hand, Columbus learned or confirmed some items of information fundamental to the elaboration of his plans for crossing the Atlantic: in the first place, that the world was a perfect sphere—an inaccurate observation but universally acknowledged as a truth at the time, which served the explorer's purpose until, as we shall see, he discarded it on the strength of his own observations in 1498. Secondly, Ptolemy taught that the known world extended in a continuous landmass from the western extremities of Europe to the easternmost limit of Asia and that between the two points lay an intervening ocean; this had been a commonplace of the medieval world-picture before the reception of Ptolemy; but his authority confirmed that it was theoretically possible to pass from Europe to Asia across the Atlantic. The last point on which Ptolemy's lore coincided with Columbus's plans was that to the south of the known world existed unknown lands: depending on the latitude he chose for his crossings, this left open the possibility of making substantial new discoveries. Moreover Ptolemy—as most readers understood him—blocked the eastern route to India with hypothetical lands, enclosing the Indian Ocean. There is no evidence that Columbus shared this view but it was taken seriously at the time: in 1490 Portugal sent an espionage mission into the Indian Ocean precisely in òrder to verify it. If the route around Africa was inaccessible, that would be an additional reason for attempting the direct transnavigation of the Atlantic instead.

Ptolemy offered encouragement to Columbus the cartographer as well as to Columbus the explorer. Columbus adopted the Alexandrian's principle of constructing maps on a grid and fixing the positions of places by co-ordinates of longitude and latitude. The map recording his Atlantic discoveries, promised for his patrons, was conceived on Ptolemaic principles, and the effort—always wildly unsuccessful—to determine longitude and latitude punctuates his accounts of his voyages. As no reliably genuine map from Columbus's hand survives, it is impossible to be sure how faithful he was in practice to Ptolemy's standards. The history of the making of his map of the New World is represented only by the Spanish monarchs' impatient demands to see it finished—which raises the possibility that it always remained incomplete. It has been argued that the map most likely to have been modelled on Columbus's own work—a Turkish map of 1513, compiled from captured Spanish documents—does show signs of having been copied from a chart laid out on a grid;[34] but as Columbus was a poor reader of latitude and, like all his contemporaries and those who followed for more than a century, never remotely approached a solution to the problem of determining longitude, any efforts he may have made are bound to have been tentative, at best.

In other ways, Ptolemy's doctrines were less serviceable for Columbus. In Ptolemy's opinion, and it was shared by most of Columbus's contemporaries, the known world occupied exactly half the circumference of the globe. Though Ptolemy allowed that the unknown Orient might extend beyond that point, an Atlantic crossing would involve a journey across half the breadth of the globe—an amount that must have equalled a distance beyond the reach of any vessel of the day, especially on the relatively southerly latitudes Columbus was to choose. Moreover, Ptolemy's calculation of the length of the earth's circumference was too generous for Columbus's taste—about 8 per cent greater than the figure the explorer espoused. Columbus responded by rejecting Ptolemy and searching for alternative authorities who promised a shorter journey. Ptolemy himself introduced him to the first of them, Marinus of Tyre, who exceeded Ptolemy's estimate of the extent of the world's known landmass by forty-five degrees. His figure had survived only because of Ptolemy's dismissal of it. Towards the end of his life, Columbus would claim that he had proved Marinus right and Ptolemy wrong.[35]

His attitude to Ptolemy is a curious indication of how his mind worked and of the problems of scientific investigation at a time when experiment was beginning to rival tradition as a source of scientific authority. Columbus had a profound respect for the texts: as would be natural for a self-educated man, he probably felt a certain awe in their presence. But he knew that they could not satisfy his longing to know 'the secrets of this world'; later, from his own experience, whenever he was able to disprove something Ptolemy had said, he would gleefully exult. He had already taken pride in being a witness to the fact that the tropic zones were habitable, *pace* the sage of Alexandria, on his voyage to the Gold Coast (even though, like so many of Columbus's observations, this was vitiated by error, since at the time he supposed himself on the Equator he was actually five degrees to the north of it).[36] On the other hand, the study and knowledge of the texts and the acceptance of the authority even of Ptolemy, when it lent itself to Columbus's purposes, were seminal influences in the emergence of his ideas. Because Columbus's notes on Ptolemy's *Geography* have not survived, it cannot be quarried, like the other annotated books, for particular insights into Columbus's values and priorities and the workings of his mind. It does seem fair, however, to assign it a primordial place in the formation of his notions of geography.

A partial corrective to Ptolemy was to be found in the travels to the limits of Asia described by *The Book of Marco Polo*. Columbus's edition, published in 1485, was almost certainly not acquired for his library until 1496, but his own writings show that he was well acquainted with Marco Polo's version of Oriental place-names by 1492. The traveller's text was very old and much perused by Columbus's time, but its authority was a matter of controversy. Among scholars in Italy and southern Germany the Venetian was most cordially trusted, whereas in other areas and among more traditional scholars his book was regarded with scepticism. In Spain it seems to have been little known. Medieval men had been taken in by too many fables of untold wealth and unseen prodigies of nature in the East to give ready credence to a tale as full of marvels as Marco Polo's. The book's common soubriquet, *Il milione*, is an ironic tribute to the hyperbole of a supposed charlatan. The text did not confer the sort of authority Columbus would have been well advised to invoke when arguing the merits of his plan; but he was always uncritical in his

selection of evidence and found Marco Polo especially serviceable in
three ways. To begin with, Columbus reckoned that the Venetian's
travels in Asia must have taken him well beyond what Ptolemy had
reckoned the furthest limit of the land. This in itself would whittle down
Ptolemy's unnavigably large ocean. Moreover, Columbus made a
special note of Marco Polo's report of no less than 1,378 islands off the
coast of Asia. This amounted to a promise of a landfall before the
completion of the crossing to the mainland. Finally, Marco Polo had
reported, 1,500 miles out from China, the gilded, gardened, watered
island of Cipangu. This was the first, and remarkably reliable, notice to
reach Europe of the existence of Japan, but because it was uncor-
roborated, it was a matter of doubt. Marco Polo had misreckoned its
distance from China and given no adequate indication of its whereabouts. Nevertheless, Columbus snatched at Cipangu like a golden
straw in the midst of the ocean. On his first Atlantic crossing, though he
did not initially steer for it he altered course in the hope of finding it.
While in the Caribbean he sought it frequently and sometimes thought
he had found it. The early tradition that his suit was dismissed at the
Portuguese court because he was 'fantastic in his imaginations with his
island of Cipangu' reflects a reputation tainted by contagion with Marco
Polo.[37]

Columbus's interest in the Venetian traveller was, at best, impurely
scientific. He was drawn by the exoticism and extravagance of the
wonders of the East. His annotations show no interest in geography or
ethnography and much in the wealth of Eastern realms. In the pages of
Columbus's copy of Marco Polo we are closer, perhaps, to the explorer's
literary tastes than to his geographical theories. Marco Polo was a
merchant by birth and a functionary by adoption. The travels he
described were undertaken in the service of Kublai Khan, who commanded him to provide entertaining reports of his observations, rather
like a male Scheherazade. It was as a story-teller that the Venetian
excelled, and as a titillator of his audience. We can think of other reasons
than objective curiosity for his descriptions of Tibetan sexual hospitality
or his practised evocations of the trade-tricks of Chinese whores. His
candid assurances of the existence of tailed men, men with dogs' heads,
and islands respectively of males and females who came together
periodically to breed all justified his reputation as a mere travelling
fable-monger. At least Marco Polo really did travel among some of the
lands he described so sensationally. Another travel-writer who captured
Columbus's fancy was the notorious 'Sir John Mandeville', who in-

dulged in fancies much richer than Marco Polo's without travelling further than the nearest book-cupboard. What particularly attracted Columbus, apart from the amazing stories, were the lists of exotic products he noted in the margins of his copy: 'spices, pearls, precious gems, cloth of gold, marble', ginger, sugar, silks, mines of azure and silver, houses smothered in gold, copious victuals, and abundance of rich merchandise.[38]

At first sight, one might hope for more help with the sources of influence on Columbus's geographical thought from his copy of Pierre d'Ailly's *Imago Mundi*. D'Ailly was the most heavily perused author in the explorer's library. Fragments of his book, and of two cosmographical and astrological treatises bound with it, were plucked from their context, memorized, strung together in astounding patterns and regurgitated in support of some of Columbus's most contentious—even bizarre—later theories: such as, from 1492 onwards, that he had discovered a short route to Asia, or, in 1498, that he had located the earthly Paradise, or from about 1500, that his discoveries were divinely ordained as harbingers of the millennium. From d'Ailly's pages Columbus got some of his speculations about the existence of the Antipodes and most of his arguments in favour of a small world and a narrow Atlantic, including al-Farghani's figure for the length of a degree; from the same source, he stored up annotations which reveal an interest in methods of prediction of the date of the millennium; and he copied a table of the length of the solar day at the solstice by latitude, which, as we shall see, he used on his first transatlantic voyage as the basis of his attempts to record his latitude as he sailed.[39] So vital is d'Ailly's influence that it seems particularly important to establish the date at which it was exercised. Columbus's edition was undated, but published in 1480 or 1483; not even this fact establishes a firm *terminus a quo*, since Columbus may previously have had access to an earlier text. One of the marginal annotations refers to the year 1481, as if in the present tense, but could be a quotation. An unquestionable *terminus ad quem* for Columbus's first reading of the book is provided by another annotation which refers to an event of 'this year of 1488'; at least one annotation, at a point in one of the appended treatises where d'Ailly discusses astrological methods of divining the date of the end of the world, was written as if 1489 were still in the future. A further note refers to 'the present year 1491' and treats March, 1491, as if in the future.[40] But, as the work may have been—indeed, almost certainly was—read many times over in the course of Columbus's life, as he scoured its pages for more 'secrets of this world', more clues to the nature of his discoveries, more arguments in favour of his own claims, it is impossible to draw any reliable conclusions about the precise chronology of the evolution of Columbus's ideas. That in the 1480s he was already contemplating with interest the prospect of the end of the world, does not mean, for instance, that he was already imagining the personal role which he later ascribed to himself in precipitating that consummation devoutly to be wished. D'Ailly's book provides not so much a guide to the development of Columbus's thought as a window on to the range of his priorities. For the most vivid general impression conjured by his annotations—underlying all the particular instances of his concern with geographical problems, with Atlantic projects, and with astrological prognostications—is his abundant love of the exotic. The most heavily annotated part of the book is saturated with images of the marvels of the East and the riches of India—in gold and silver, pearls and gems, fauna and fabulous beasts.

The same picture of a Columbus attracted by the exotic and excited by wealth emerges from his annotations to Pius II's geographical compendium, the *Historia Rerum ubique Gestarum*. Columbus's copy, printed in 1477, was almost as heavily annotated as his *Imago Mundi*, with 861 postils to the latter's 898. The margins of both works are speckled with drawings of little fists with extended index fingers pointing to passages of special interest or curiosity. They cover a huge field, but are overwhelmingly concerned in one form or another with the riches and diversity of the Orient. Apart from the Amazons, hydrography, and general exotica, the specific topics in the book which most engaged Columbus's attention were the navigability of all oceans, the habitability of all climes,[41] and the question of the existence of the Antipodes. In discussing the first of these questions, Pius II demonstrated an implicit belief—or disposition to believe—in a navigable route between Asia and Europe via the Atlantic. Columbus noted, for instance, his story of Indian merchants reportedly come ashore in Germany in the twelfth century. The humanist Pope had a habit of juxtaposing textual with empirical evidence—of using, that is, the practical results of reported navigations, the observed evidence of real journeys, to confirm or disprove the assertions of received wisdom. Columbus, who lacked formal education but laid claim to vast practical experience at sea, rested his own challenges to scholarly authority on the basis of his superior craft lore, albeit deploying written authority— increasingly, it appears, as time went on—in an ancillary role. His method may have been inspired by Pius II's example, or may have been espoused *faute de mieux*. It gives a speciously scientific character to his own writings, with their frequent appeals to empirical values. At the very least, Pius II can be said to have encouraged him to see geography as an exciting terrain of new discovery, in which few parts of the received picture were beyond cavil and a whole world was, as it were, up for challenge.

The date of Columbus's reading of the book cannot be assigned with any more confidence than in the case of the *Imago Mundi*. An annotation which refers to José Vizinho's readings of latitude along the West African coast must date from after 1485; another—which contains a further reference to Bartolomeu Dias, the discoverer of the Cape of Good Hope—must date from at least 1488. What seem to be convincing allusions to points made by Pius II in Columbus's so-called journal of his first transatlantic voyage make it virtually certain that the book was already well known to the explorer before 1492; and parts of it at least were again fresh in his mind when he made his last voyage in 1502–4.[42]

The other surviving annotated books from Columbus's library were Pliny's *Natural History* and Plutarch's *Lives*. They are late editions, of 1489 and 1491 respectively, but still available to Columbus before he crossed the Atlantic. Their influence on his planning seems negligible. Although sixteenth-century tradition gave Pliny a formative place in the making of Columbus's plans, the discoverer's only reference to him in a theoretical context is irrelevant, and almost all the twenty-four annotations are attempts to translate the Italian text of Columbus's edition into Spanish. One note, which refers to Hispaniola, must have been written after 1492. If Columbus was interested in remedies for eye complaints on his own behalf, at least one reading can be dated to after 1494, when his chronic and painful eye condition first broke out during his exploration of Cuba. Apart from cures—mainly for gallstones—Columbus's interests were chiefly engaged by the same range of subjects noted in his copies of *Marco Polo*: gold, silver, pearls, amber, and 'manifold marvels'.[43]

In the Plutarch, which is annotated in greater detail, it is possible to pick out more of Columbus's particular obsessions. Of 437 notes, ninety-nine are concerned with auguries, portents, and occasionally more recondite forms of divination, like Numa's conjuration of demons. Within this category, Columbus shows himself particularly interested in visions, and takes special note of any 'voices in the air', like those which counselled Marcus Caecius. It is not hard to detect here evidence of an interest which might have been aroused by his own dialogues

with his 'voices'. With exceptional broad-mindedness, Columbus also
noted Dio Cassius's case against believing in visions. The next best-
represented theme is that of deceit as an instrument of policy:
Columbus noted every instance of a politician's ruse, or a commander's
stratagem. The captain who tampered with the compass and faked the
log was bound to find comfort in these examples. Columbus also took an
exceptional degree of interest in instances of extreme phlegmatism and
sang-froid épatant—particularly in heroes like Brutus and Manlius
Torquatus, who were willing to exact the ultimate penalty of justice
from their own sons when duty so required. Columbus never made the
parallel explicit, but it is worth recalling that in Hispaniola in 1500 he
felt reluctantly obliged to put a mutineer to death: in justifying this
arbitrary and fatal proceeding, he declared that he would not hesitate to
do the same to his own son in similar circumstances. He admired the
dispassion of Pericles, whom nothing could move to tears save his son's
plight: Columbus could sympathize with the Athenian, for he had been
moved to feelings of intense suffering for his own boys—especially
during his return voyage from his first Atlantic crossing when he
thought he would never see them again, and during his last when he
beheld the suffering and fortitude of his younger son, who accompanied
him. His annotations also dwell on types of 'good death' and, more
generally, on how the heroes of antiquity died. And some of his familiar
obsessions are represented by odd jottings: a *femina fortis* is noted in the
form of Manlius's wife; the whereabouts of the Amazons are raised.
Odd anecdotes and moral tales reveal an aspect of Columbus's familiar
taste: he relished, for example, the story of the theft of a lion-cub who
ate his raptor, or of two women of Rome who died of pleasure, or of
Numa's love of eagles, or of the prodigious chameleon who can assume
every colour except white. The punctiliousness with which he made a
note of heroes susceptible to 'the unspeakable lust' makes it surprising
that no one has yet written a book claiming that Columbus was gay.[44]

Thus, although it is difficult, perhaps impossible, to establish a
coherent account of Columbus's geography from the evidence of his
reading habits, it is possible to glean some insights into his tastes. He
was bookish but not scholarly; a 'reading man' whose inclinations were
lowbrow. He liked the sensational and the trivial, the sententious and
the salacious. He picked up points which echoed his own experience or
which related to his own ambitions. From his annotations, those
ambitions emerge as material, at least as much as scientific: he was
interested in Asia for its yellow-press 'marvels' and golden-book wealth.

His attitude to scientific authority was a curious mixture of the servile and the reactive. He garnered nuts of useful information like a squirrel, and cracked them like a critic. From the sparse evidence which survives, it is tempting to see his interest in the spoils of scholarship growing as time went on and to suppose that his reading grew more purposeful from 1498 when he plundered his books systematically for a defence of his career so far; but that does not necessarily mean that he had not attempted an academic exposition of his ideas before 1492.

By 1492, indeed, Columbus had acquired enough book-learning to add the attributes of an amateur geographer to the accomplishments of a seasoned navigator. The indefeasible problems of the chronology prevent us, however, from being confidently specific about the geographical theories he advocated and the project he based on them. The marginal notes in his books are the only evidence we have of what was going on in his mind before 1492, and even of these many or most may have been written later. That he was contemplating a transnavigation of the Atlantic is confirmed by the notes; so is the proposal that he was interested in Asia, but from what these sources tell us, it is not possible to say when the Atlantic project became identified with the quest for Asia in his mind. Before 1492 he had also contemplated a voyage to discover the Antipodes; on the basis of some of the evidence, he might have been content simply with the discovery of new islands. As he wavered between three possible objectives, he seemed not so much particular about what he proposed to discover as insistent on his determination to discover something. Against an awareness of the high—perhaps overriding—place given to social self-advancement in Columbus's scale of ambitions, the unfixed and unfocused nature of his evolving geography is easy to understand. He may have cared less about where he was going than about whether—in a social sense—he would 'arrive'.

By 1492 he had still not quite committed himself to more than a search for 'islands and mainlands'; but in the course of that year, all the emphasis that can be discerned in the sources came to be placed on the hope of finding a short route to the East. This emerged both from the preparations Columbus made—shipping an Oriental interpreter, carrying passports implicitly addressed to the ruler of China, assuring his patrons that he would 'go to the east by way of the west—and from the consistent evidence of his own account of his first Atlantic crossing, which insists on an Asiatic destination with convincing monotony. From the time he set sail Columbus never mentioned any other possible destination and when he returned, though pundits were not lacking to identify his new discoveries as the Antipodes or simply as previously undiscovered islands, Columbus rigorously excluded all but Asiatic descriptions from the way he presented them. If the evidence of his intellectual formation yields no help with the problem of when and how this narrowing of focus came about, it may be useful to look for assistance to the other process in which Columbus was engaged in the same years—the search for a patron.

3

'His Manifest Hand'

THE QUEST FOR PATRONAGE
*c.*1484–1492

T w o episodes of the Columbus legend contrast the footsore penury of
the friendless projector, who arrived penniless in Spain, with the
crowning triumph of the moment in 1492, 'six or seven years' later,[1]
when, in Columbus's own words:

on the second day of the month of January, I saw the royal banner of your
Highnesses raised, by force of arms, on the towers of the Alhambra, which is the
fortress of the said city, and saw the Moorish king come to the gates of the said
city and kiss the royal hands of your Highnesses and of my lord the Prince . . .
and thereafter, in that same month . . . your Highnesses, as Catholic Christians
and as princes who love the holy Christian faith, and as augmenters thereof and
foes of the sect of Muhammad and of all idolatries and heresies, thought of
sending me, Christopher Columbus, to the regions of India . . . and your
Highnesses ordered that I should not travel overland to the east, as is customary,
but rather by way of the west, whither to this day, as far as we can know for
certain, no man has ever gone before.[2]

Neither of these episodes may have been quite as the legend represents
them, but the problem they pose is real enough: how did Columbus,
who by his own admission had nothing to offer except promises which
'were neither few nor vain',[3] obtain the royal sponsorship which
launched his enterprise and his career?

Fifteenth-century explorers, as far as we know, did not set off into the
blue without the licence and support of some mighty prince. Columbus
explained clearly enough why this was so: a private individual might
make a discovery, but could not advance a claim to sovereignty; nor
would he be able to keep his gains without princely protection against

interloping by fellow-subjects or attack from some foreign prince. When Columbus was accused in the late 1490s of plotting to alienate his discoveries from the Castilian Crown, his scarcely relevant defence was that he could not do without a patron. Only a state had the authority to legitimate his enterprise and the power to deter poachers. As a Genoese, Columbus was free to seek state patronage far and wide: he seems, at various times, to have contemplated his native republic, the Pope, and the monarchs of Portugal, Castile, France, and England as potential protectors. In his own warped memory, it came to seem as if he had fended off eager rivals for his services and that God gave the project to Castile, against bids from the English, French, and Portuguese, 'by a miracle', procured by God's 'manifest hand'.[4] In reality, Columbus excited only intermittent flickers of interest outside Spain, and the sponsorship of Ferdinand and Isabella was won by dint of long and unremitting effort.

The tradition that the search for sponsorship began in Portugal in 1484 can be regarded as provisionally reliable, though it is worth bearing in mind that no direct evidence survives of an approach by Columbus to the court of Portugal before 1488, when he had already removed to Castile.[5] Portugal, certainly, was no bad place to start. All known new discoveries in the Atlantic since the middle of the century had been made under Portuguese auspices, and the experience of Portugal's war on the high seas against Castile in 1474–9 had demonstrated, on the whole, that the King of Portugal could deliver a patron's best gift to an explorer: protection of his discoveries against the pretensions of rivals. The incumbent King, moreover, João II, was personally committed to the development of Portuguese exploration and to the extension of his kingdom's reach along the supposed route to the Indian Ocean down Africa's west coast. He tried to give the entire African enterprise enhanced prestige at home. He took the title of 'Lord of Guinea'. He emphasized Portugal's claims to sovereignty in Africa—doubtless with a wary eye on Castilian envy—and the duty of evangelization which was thought to legitimize it. He presided over an extraordinary 'turnover' in baptisms and rebaptisms of Black chiefs. He built the impressive trading establishment of São Jorge da Mina near the mouth of the Volta to boost the gold trade, and he centralized African commerce in Lisbon, in the Casa da Mina, beneath the royal palace, where all sailings had to be registered and all cargoes warehoused. It became Columbus's habit to taunt the Castilian monarchs with the energy and commitment of this King.[6]

Against this background, if Columbus did seek support in Portugal for an attempted Atlantic crossing in 1484, it is surprising that it was rejected. Only three years later a Flemish adventurer, Ferdinand van Olmen, received a commission from Portugal for what seems to have been a closely similar project to find 'islands and mainland' in the Atlantic Ocean.[7] To judge from the frequency with which such documents occur in Portuguese archives, commissions to explore were easy to come by: eight survive from between 1462 and 1487.[8] Yet according to the traditions on which we are compelled to rely, Columbus's suit failed for two reasons. The committee of professional savants who sat in judgement on his plans rejected them because they disbelieved in the existence of Cipangu; and Columbus in any case demanded excessive rewards for his services in the event of success.[9] Both explanations are credible: Columbus's attempts to obtain King João's support might equally well have been vitiated by the improbability of his plans or the importunity of his demands. Other possible, but entirely speculative reasons are that Columbus was unable to raise private capital for his undertaking or that departure from the Canary Islands was already an integral part of his plans—for the Canaries were a sphere of Castilian expansion, into which the Portuguese, despite repeated efforts, had been unable to break. Columbus's own explanation, advanced later, with the advantages of hindsight and the exaltation of a divine conviction, was that God had occluded the King of Portugal's vision in order to reserve the glory of the discovery for Castile.[10]

Divine plans mature slowly. In this case, certainly, God seems to have been in no hurry. When, from his disappointment in Portugal, Columbus transferred his frustrations to Castile, he still had those six or seven years of waiting ahead of him, checkered with moments of despair, before he at last obtained a royal commission. Castile's record in promoting Atlantic discovery, though a long one, was patchy and fitful. Castile had lagged behind Portugal in the competition for territories overseas, for want not of will but of means. The Castilian kings had disputed the claims of Portugal in Africa and the Atlantic islands since 1345, when both Crowns had claimed the right of conquest of the Canaries before the court of Pope Clement VI. On that occasion, the Pope had pre-empted both rivals by giving away the perquisites himself. From early in the fifteenth century, Castilian jurists had evolved an argument which attributed rights of conquest in Africa to their own sovereign, by virtue of the supposed descent of the regality of the Visigothic rulers of all Spain, together with their rights against the

Moors, through the line of the monarchs of Castile. Little of practical consequence had been achieved, except for the conquest of four of the Canary Islands. By the time Columbus arrived, however, the pace of Castilian involvement was quickening.

Castilian interlopers in the African trade had attracted Portuguese complaints since the 1440s, but the war of 1474–9, in which Ferdinand and his wife Isabella were challenged from Portugal for the Crown of Castile, acted as a catalyst for Castilian activity. The monarchs were open-handed with licences for voyages of piracy or carriage of contraband. The Genoese of Seville and Cadiz were keen to invest in these enterprises, and Andalusian mariners, including many who were to ship with Columbus or who made transatlantic journeys after him, were schooled in Atlantic navigation. The main action of the war took place on land in northern Castile but was accompanied by a 'small war' at sea in the latitudes of the Canaries. Castilian privateers were licensed to break into Portugal's monopoly of the Guinea trade by force. The Genoese governor of the Portuguese Cape Verde Islands, Antonio da Noli, defected to Castile. Portuguese ships made numerous attacks on Castilian settlers in the island of Lanzarote. The importance of the unconquered islands of the archipelago—precisely the richest, Gran Canaria, La Palma, and Tenerife, which were still in the hands of their aboriginal inhabitants—and the fragility of the Castilian hold on other lands were thrown into prominence. When Ferdinand and Isabella sent a force to resume the conquest of the Canaries in 1478, a rival Portuguese squadron was already on its way.

Meanwhile other, longer-maturing reasons coaxed the Castilian monarchs into an Atlantic policy. The Portuguese were not the only rivals for possession of the Canary Islands: the title of lord of the islands had descended to Diego de Herrera, a minor nobleman of Seville, who fancied himself as a conquistador. He was typical of the sort of truculent paladin whose power in a peripheral region was an affront to the crown. Profiting from a local rebellion against seigneurial authority in Lanzarote in 1475–6—one of a series of such rebellions—the monarchs determined to enforce their suzerainty. In November 1476 they initiated an inquiry into the juridical basis of the lordship of the Canaries. Its findings were embodied in an agreement between seigneur and suzerains in October 1477: The Herreras' rights were unimpeachable, saving the superior lordship of the Crown, but 'for certain just and reasonable causes' which were never specified, the right of conquest reverted to the Crown. Between 1480 and 1483 Gran Canaria was laboriously conquered against natives who, armed literally with sticks and stones, exploited the harsh terrain to win repeated victories over technically superior adversaries. Meanwhile recurrent insurgency by the natives of Gomera compelled the royal forces to be sent there from Gran Canaria: in 1488 and 1489 brutal incursions crushed the rebels who with dubious legality were enslaved in droves as 'rebels against their natural lords'. This definitive conquest of Gomera put, incidentally, the island's deep-water harbour of San Sebastián, on the western edge of Christendom, at Columbus's disposal.[11]

Beyond the Canaries, remoter Atlantic prizes beckoned the Castilian monarchs. As always in the history of Latin involvement in the African Atlantic, gold was the spur. According to a highly privileged observer, King Ferdinand's interest in the Canaries was aroused by a desire to open communications with 'the mines of Æthiopia' (that is, Africa).[12] The conclusion of the Portuguese war effectively denied him access to the lucrative new gold-sources developed from Portugal on the under-side of the African bulge, around the mouth of the Volta, in the 1480s. This must have helped to stimulate the search for alternative sources of gold and may help to explain, for instance, the emphasis on gold in the journals of Columbus. In 1482 the Castilian monarchs turned to the conquest of Granada, the last surviving Moorish state in the Iberian peninsula: this did not mean, however, that they had lost interest in the Atlantic, only that while peace with Portugal secured the rear, they could press on with an undiminished policy of expansion on another front. The conquest of the Canaries was continued, albeit at a slow pace; and in one respect the conquest of Granada stimulated interest in exploration further afield by increasing the urgency with which new sources of gold had to be sought. The costs of the war, and the sacrifice of the traditional Granadine tributes, combined with the loss of Castilian prospects in Africa, gave Columbus's proposals growing appeal in the Castile of the 1480s and early 1490s.

Since the union of the Crowns of Aragon and Castile in 1479, to Castilian aspirations of expansion were added traditional Aragonese concerns with the eastern Mediterranean and the trade-routes to the Orient. The sense of an impending struggle with Islam, which had been gaining force gradually throughout the century, was particularly strong in Spain, land of a secular conflict with the Moors and of more recent involvement against the Turks. A lively tradition, of long standing at the Aragonese court, linked millenarianism with the ambition to rule in

Jerusalem—to make a reality of the title of 'King and Queen of Jerusalem' which Ferdinand and Isabella had inherited. In the late thirteenth and early fourteenth centuries, the prophetic writings of Arnau de Vilanova had specified an eschatological role for Aragonese kings, including the renovation of the Church, the conquest of Jerusalem, and the creation of a united world-wide empire.[13] This programme was borrowed from the twelfth-century biblical divinations of Abbot Joachim of Fiore, who was one of the most influential sources of late medieval chiliastic traditions. Joachimism was widely affected by Franciscans of the sort who were to provide some of Columbus's closest friends in Spain, and may have inspired some of his most deeply held convictions. In later writings, Columbus was to cite Joachim, albeit not from direct knowledge, and showed some awareness of Arnau. In the entourage of Ferdinand the Catholic, a revival of these millenarian traditions seems to have been under way when Columbus joined the court in the mid-1480s. The King was seen by some admirers as a potential 'Last World Emperor' who would fulfil some of Joachim's preconditions, including the conquest of Jerusalem, for the end of the world.[14]

For most advocates of this vision it was, perhaps, merely a propaganda device; but propaganda has to be credible to be effective. During his time at court, Columbus could have been exposed to enough propaganda of this sort to convice him at least that the monarchs were serious about their Hierosolomitan ambitions. He would have heard musical settings of the prophecy that Ferdinand and Isabella would conquer Jerusalem, and a song of Juan de Anchieta, which attributed to 'Scripture and to saints' the fore-vision of the monarchs crowned by the Pope before the Holy Sepulchre. In 1489 he could have witnessed the reception of a party of Franciscan custodians of the burial-place of Christ.[15] According to his later recollection, Columbus proposed, as part of his submissions to the monarchs for their sponsorship of his Atlantic voyage, that the profits should be devoted to a campaign for a crusade to Jerusalem. Throughout his subsequent career, the thought of Jerusalem never left him; it was often to be recalled, as we shall see, particularly at times of severe mental stress. If, as seems likely from the evidence of his marginal annotations, his interest in the computation of the millennium was already aroused in the 1480s, and the name of Jerusalem already had a special resonance for him, it is easy to see why he should have found the court of the 'King and Queen of Jerusalem' a particularly congenial and inspiring place to be.

On a more pedestrian level, Spanish industry, commerce, and shipping were enjoying a prosperous period which added new urgency to the search for trade-routes and exotic marts while generating capital for investment. Competition at all levels between Spain and Portugal had rarely been more intense than now. In the treaty which ended the war of 1479, the two realms partitioned future spheres of expansion: the Canary Islands, including any still undiscovered, with a portion of the African coast immediately opposite them, were to belong to Castile, while the rest of mainland Africa would be the exclusive prey of Portugal. No agreement, however, could be definitive in the unstable ambience which then prevailed, and by 1482, during marriage negotiations between the two dynasties, the terms were in the melting-pot again.[16]

It was therefore natural that in the 1480s Columbus, looking for a patron, should oscillate between Portugal and Spain. Furthermore, Ferdinand and Isabella were not the only protectors Castile had to offer; if the tradition which dates his removal to Castile to 1485 can be trusted, he seems to have spent more than his first year there, in forming slightly more modest connections. Overseas expansion was not in Castile, as it was in Portugal, a long-established 'public sector' activity, tightly controlled, and in some spheres strictly monopolized, by the Crown. It was open to any Castilian subject who possessed the means and motive to take some ships on a slaving *razzia*, capture a Berber town, trade illicitly in Portuguese Guinea, conquer a Canary island or invade the Kingdom of Granada—or, if he so wished, attempt to sail across the Atlantic. Enterprises such as Columbus proposed added the lustre of glory to the hope of commercial gain, and neither of these was beneath the dignity of a great nobleman in the Castile of the day. In particular the Count of Medinaceli had invested handsomely in mercantile marine ventures, and had a traditional family connection with Castilian expansion in the Canary Islands, while the Duke of Medina Sidonia was involved in shipping, provisioning, and sugar-trading and was interested in taking part in the conquest of the Canaries.

To these potential sources of help Columbus addressed early appeals. What he proposed, according to Medinaceli's later recollection, was a voyage 'to the Indies'. In addressing a *bourgeois gentilhomme*, Columbus seems to have stressed the most glittering prize, rather than the imponderable benefits of the Antipodes or of more new islands. All he required were 'three or four caravels, for he asked no more'. His plan was favourably received and Medinaceli later claimed to have maintained him for some time; but the nobleman seems to have felt that the project was of such magnitude as to require royal approbation. A trip to the Indies would necessarily involve far more than commercial considerations and raise the question of the sovereignty of the lands visited, with negotiations in Portugal and, no doubt, supplications to the Pope. In sending him on to the royal court, Medinaceli may also have been influenced by Columbus's threat to take his project to France: as this was a form of blackmail to which Columbus frequently resorted in his later relations with his patrons, Medinaceli's reference to it is entirely credible.[17]

In this early period of his life in Castile—if it is correct to date his association with Medinaceli so early—Columbus contracted an intimate alliance with a woman of Cordova called Beatriz Enríquez. She was the daughter of peasants and the ward of her uncle, Rodrigo Enríquez de Arana, who was a modestly well-to-do citizen. But her family connections, which included such lowly artisans as carpenters and butchers, were not what Columbus demanded in a potential wife. Columbus was not, by common standards, a sexually susceptible man. Only one other amorous encounter outside marriage was ever imputed to him. His response was markedly prudish when his followers indulged in concubinage or promiscuity with native women in the New World. Later in life, he affected the dress and habits of a coenobite and generally appeared to prefer the company of friars to that of women. The incontrovertible proof of his passion for Beatriz, however, was the birth of their son Fernando in November 1488. Fernando turned out to be a highly satisfactory boy, whom Columbus acknowledged with pride. He accompanied his father's last voyage to the New World, wrote an account of his father, and became one of the most famous men of letters of his time. Columbus formally legitimized him and commended him to his legitimate elder brother in unequivocal terms: 'Advise your brother as the elder son should the younger. He is the only brother you have and may our Lord be praised that he is just such as you need; for he has turned out to be of very great learning.'[18]

His mother's position was never regularized: marriage with a woman of such humble provenance would have compromised Columbus's hard-won status. Columbus seems, however, to have behaved responsibly and even affectionately towards her. In one of his late memoranda to his elder son, he reminded him, 'Take Beatriz Enríquez in your charge for love of me, as attentively as you would your own mother. See that she gets 10,000 maravedis a year beyond her income from her meat business in Cordova.' In a late codicil to his will he declared that his solicitude for her was 'because of very great obligations' he owed her and 'for the discharge of my conscience, for this weighs heavily upon my soul. The reason for it'—he added with an evident allusion to the irregularity of their relationship, and an equally evident touch of remorse—'cannot lawfully be written down.'[19]

It is possible that Beatriz Enríquez was not his only amorous encounter in Cordova, for he may also have met there a lady with whom he had no verifiable affair but an association which excited rumour. Beatriz de Bobadilla was one of the cruellest and most beautiful women in Castile. From an involvement with the King himself, she had gone on to the conquest of the Canaries in 1481 as the wife of the conquistador Hernán Peraza. When he was murdered by rebellious natives in 1488, she was to become mistress of the island in her own right, suppressing the rebellion bloodily and enslaving many of the islanders. She combined the qualities of *femina fortis* and *femme fatale*. She was such a good subject for scandalous tittle-tattle that every tale about her is bound to be sceptically received. The source, moreover, of the story that linked her to Columbus was the notoriously prurient Michele de Cuneo of Savona, who accompanied the second crossing of the Atlantic and whose other stories include a particularly graphic account of his own seduction, with the aid of a length of rope, of a native girl. Still, Cuneo mentions the affair in so offhand a manner as to be convincing. Relating the expedition's halt in Gomera, 'If I were to recount for you', he says, 'how many festivities, salvoes and salutes we performed in that place, it would take too long; and it was all for the sake of the lady of the said place, for whom in a former time our Admiral had been smitten with love.'[20]

Between May 1486 and September 1487 Columbus was maintained in Castile, probably for most of the time at court, at the monarchs' expense. It was during this period—perhaps, to judge from the availability of the experts called to hear his case, in late 1486 and early 1487—that his Atlantic project was submitted to the scrutiny of a panel of 'wise men, learned officials and mariners' commissioned by the monarchs to evaluate it.[21] This episode has generated an enormous amount of speculation and a great deal of incredible legend—including the infamous canard that the 'experts' thought the world was flat.[22] The verifiable facts, however, are few. Only two members of the panel are known by name: its chairman, Fray Hernando de Talavera (the monarchs' Jeronimite confessor who later became Archbishop of Granada) and Rodrigo Maldonado de Talavera, a former professor of law in the University of Salamanca, who joined the royal Council as a full-time administrator in 1480. Maldonado's later recollection summarizes the outcome: 'All agreed that what the Admiral was saying could not possibly be true, and against their opinion the Admiral determined to go on the said voyage.'[23] An early tradition transmitted by Las Casas implies, as we have seen, that the question at issue was not—or not only—a short route to Asia, but the existence of the Antipodes. Mariners who later testified, in another connection, in a trial between the Columbus family and the Crown, recalled what may have been another of the panel's scruples: that Portuguese searches had revealed no new land in the Western Ocean.[24] No other evidence is close enough to the event to be worth considering.

The outcome of this encounter with the experts can hardly have been encouraging to Columbus and helps to explain why he should, for a while, have returned to resume his efforts in Portugal in 1488.[25] It was by no means, however, a definitive failure. Our heroic impression of him, enshrined in a historical tradition which has always been susceptible to the romantic possibilities of the Columbus story, may be misleading. It makes rattling good history to accept Columbus's self-evaluation as a despairing outcast, driven by derision, but it may be more convincing to stress the positive aspects of Columbus's predicament—the sources of momentum that carried him beyond the crisis of the commission's report to ultimate success. He was to become the author of his own legend, in which he waged a lone, long and resolute struggle, in ever-adverse circumstances, to a final triumph against the odds. Yet new notes are rarely struck by one-man bands, and the objective scrutineer of the Castilian court in the late 1480s will see a Columbus surrounded by friends and helpers. It is true that his idiosyncratic cosmography was respected by few experts; but experts rarely rule, and it was on political and financial backing, not informed assent, that the success of his quest for patronage depended. Gradually, between 1486 and 1492, with increasing force—it seems—from 1489, Columbus built up in his support a head of pressure which became irresistible. He acquired lobbyists who clamoured in high places and power-brokers whose political authority was more than an adequate counterweight to the scepticism of the savants; and with rather more difficulty he acquired a following of potential financial 'angels' whose willingness to invest in the enterprise made the decisive difference.

Many elements of the context contributed to eventual success:

Columbus's own suit at court was waged with persistence, we know, and, we can suppose, with his customary loquacity—capable of boring into submission those whom it could not persuade. He got material help from the prominence and sympathy of many of his fellow-Genoese, supplemented by a group of Florentines. The growth of confidence of a purposeful political leadership in and around the persons of the Catholic monarchs stimulated invention of and investment in expansionist projects. Rivalry between Castile and Portugal sharpened Castile's interest in overseas conquests and access to exotic trade; and in some respects, the war against Granada on which Ferdinand and Isabella embarked in 1482 favoured Columbus's design, because, although it distracted the monarchs and absorbed some of their energies, it created an urgent need for new gold-sources to replace those forfeit by the loss of Granadine tribute. The stimulus which united all these elements was the conquest of the Canary Islands, which brought together a group of administrators and financiers who were to become the kernel of the Columbus lobby.

At the nerve-centre of the monarchs' war-effort in the Canaries, scraping contingents together, assembling groups of investors, devising financial expedients, was Alonso de Quintanilla, a treasury official who was one of the most influential architects of policy in the reign of Ferdinand and Isabella. He seems to have been given responsibility for the organization of the conquest from 1480, when dwindling returns from the sale of indulgences caused a crisis of finance. He thought up a wide range of measures, including the mortgaging of royal booty and recourse to Italian, chiefly Genoese, capitalists. In doing so, he adumbrated the circle that would later contribute to the financing of Columbus. Quintanilla himself was instrumental in arranging the backing for the 'enterprise of the Indies' as for that of the Canaries. In both cases, he was supported by key figures, the Genoese merchants of Seville, Francesco Pinelli and Francesco da Rivarolo. Pinelli had been involved in Canarian finance for as long as Quintanilla, as he had administered the receipts from the sale of indulgences for the conquest from March 1480. Quintanilla's first personal subvention to the cause was made in April of the same year. Pinelli went on to acquire the first sugar-mill on Gran Canaria and make loans to the conquerors of other islands of the Canaries. For his part as a backer of Columbus, the monarchs made him one of the first administrators of the New World trade when it was organized as a royal monopoly in 1497. Francesco da Rivarolo may have done even better out of the whole affair. There is no evidence that he contributed personally to the conquest of Gran Canaria, but his son-in-law was one of the biggest investors, and the Rivarolo family was a close-knit business partnership carefully managed by the patriarchal Francesco. He took part in the financing of the conquests of La Palma and Tenerife in his own right and became the richest merchant in the archipelago, with interests essentially, but not exclusively, centred in sugar and dyestuffs. He was a mainstay of Columbus, whose fourth voyage he helped to finance and whose shipping interests he helped to manage in the last years of the discoverer's life. Some non-Genoese of Seville from close to the centre of Columbus's world also took a hand in paying for the conquest of the Canaries: the Duke of Medina Sidonia, whom Columbus (according to sixteenth-century tradition) saw as a possible patron; the Florentine Gianotto Berardi, who probably advanced part of Columbus's personal stake in the first transatlantic voyage. There seems to have been sufficient overlap for the conquest of the Canaries and the discovery of America to be seen, to some extent, as the work of the same group of men.[26]

The second focus of Columbus's support at court was in the entourage of the heir to the throne, the Prince Don Juan. He was the idol of an extraordinary little court of his own within the royal household. His entourage was not formally constituted as a separate court until 1486, but his servants, supervisors, and companions constituted a distinct group from his infancy in the late 1470s. Head and members together, they formed an odd body: a dull, weak boy, attended by some of the most distinguished and powerful men and women in Castile, amid a precious ritual. The atmosphere of the Prince's circle was captured in a description left by one of his attendants, Gonzalo Fernández de Oviedo, future historian of the Indies.[27] The personnel of the Prince's train was brilliant and his service was a ladder of promotion to positions of power and influence in court and kingdom. Among the administrative staff were some of the most promising of the lettered men, such as Gonzalo de Baeza, treasurer of the Prince's court, who was later advanced to a corresponding position in the household of the Queen; Juan Velázquez de Cuellar, the keeper of the accounts, who went on to do the same job in the royal household; and Juan de Cabrero, the Prince's chamberlain, who was promoted to serve the King in the same capacity—much to Columbus's future advantage. These officials were no mere cyphers but powerful men in their own right, key links in the chain of patronage by which the monarchs sought to increase their power. Opportunities for exploitation of patronage within the Prince's household were limited, for its personnel was paid directly by a royal secretary. Even so, as Oviedo remarked, 'The treasurers can bring profit to many in the exercise of their office', which gave direct access to the King and Queen, increasing as their careers made progress.

Alongside the officials were the Prince's 'companions': those of his own generation as companions and entertainers, a group joined by Columbus's son Diego in 1492,[28] and a body of older men for his surveillance and improvement. The younger set included Antonio de Torres, future Governor of Gran Canaria and comrade of Columbus, and the older included Nicolás de Ovando, future Governor of Hispaniola. A permanent tutelary role was exercised by the Prince's tutor, Fray Diego Deza, of the Dominican Order, who left his charge, in Oviedo's words, 'very well learned in all that which was proper to his royal person; especially was he a very Catholic and great Christian.' The Prince's mind, however, was shallow and inelastic and this was the only part of his studies for which he showed much aptitude. Deza's work was assisted by the whole atmosphere of Don Juan's court:

In the time of the Prince, my master, at his table and in his closet, in his kitchen or at his cup or buttery, or in any office whatsoever that was exercised anywhere in the palace from its very threshold, there was no room for any man that was not of pure lineage, a nobleman of refined and unmixed blood, or at the very least from a family which had always been Christian save for two or three whom I prefer not to name and whom the Queen had appointed before the Prince had a household and accounts of his own; and even these were well known to be strangers to the Prince, out of his grace and favour.[29]

Oviedo also claimed, with less emphasis, that Juan emerged 'a good Latinist', but the truth is that he was incapable of conversing in Latin or of any intellectually demanding activity. He was more at home risking small bets at games of chance or sharing his hairdresser's jokes than in serious study. Still, his bathroom furniture did include a chess set and it must be presumed that he was not averse to a little mental exercise while gaining physical relief. He was childish long beyond his boyhood: this reflected his need for security in an atmosphere redolent with responsibility and expectations of greatness which far exceeded his modest capacities. He never slept without a nightlight. He had an insatiable sweet tooth. His closet was always stocked with sweetmeats for him to suck, especially fruit preserves, stiff quince jelly from Valencia, ethereal concoctions of egg-yolk and sugar and aniseed balls. This may in part have been an inherited taste, for his parents are known to have gorged

themselves on sweetmeats on at least one royal visit to Valencia. All the
monarchs' children seem to have been brought up on syrup flavoured
with roses and the quantities consumed were prominent in Isabella's
household accounts: the Prince was capable annually of downing syrup
of sufficient value to keep a soldier in arms for a year. Juan's abnormal
greediness for confections accords with his other infantile traits. His
problems were aggravated—it is tempting to guess—by Isabella's
queenly responsibilities, which separated him from motherly affection.
His fondness for his nurse, Columbus's confidante Juana de Ávila, was
exaggerated. 'You must have me for a husband more than anyone else',
he wrote her in a typical letter.[30]

Around this dull boy revolved a glittering courtly ritual, like bright
rays around a spent sun. His daily round began when three chamber
servants called to help him dress and wash in two silver basins. A groom
of arms buckled on the princely sword and dagger. His barber and
shoemaker were called in and 'coined absurdities'. Juan attended to his
prayers, heard mass, and shut himself in for lessons with Diego Deza.
When there was no court event or hunt on, his recreation was a modest
flutter at the gaming-table or a distribution of alms to paupers and
petitioners and tips to tradesmen and menials. His biggest expenditure
was on clothes. At his evening toilette, when he washed his hands, the
water was poured on by Cabrero or his successor, Juan de Calatayud, or,
if present, one of the grandees of Castile in the unalterable order of
precedence: the Lord High Constable; the Admiral of Castile; the Duke
of Medina Sidonia; the Duke of the Infantado; the Marquess of Villena;
the Count of Benavente. At last he would deal with any petitions or
memoranda of the day while undressing.

It is not clear what disposed members of the Prince's court particularly to favour Columbus; unlike the financiers of the conquest of
the Canaries, they had no obvious interest in an Atlantic project and it is
tempting to suppose that some unknown personal links may have been
responsible. It is certain, however, that the Prince's entourage supplied
some of Columbus's most tenacious friends. Outstanding among them
was Fray Diego Deza, the Prince's tutor, who appears from Columbus's
later letters to have been for a long time on terms of the most
extraordinary intimacy with the discoverer. In his later years Columbus
recalled with nostalgia time spent dwelling in Deza's house, referred to
the 'brotherly love' they had borne one another, and assured his son of
their mutual confidence. A surviving letter from Columbus to Deza
conveys how total that confidence was, for Columbus permits himself

direct criticism of the King, whom he expressly accuses of bad faith, to appeal against which, he complains, would be 'to plough into the wind'. It seems the sort of dangerous charge one would confide only in secret, perhaps in the privileged relationship of penitent and confessor. Deza's intimacy was of enormous help to Columbus as the Dominican rose in the service of Church and State: tutor to the Prince, 1486; Bishop of Zamora, 1494, and subsequently of Palencia, Salamanca, and Jaén; Inquisitor-General, 1499; Archbishop of Seville, 1505. Columbus ascribed to Deza the credit for having ensured that his discoveries were made in Castilian service. 'He was the cause of their Highnesses' leaving the Indies, and of my remaining in Castile when I was about to leave the country.'[31]

Other members of the Prince's household were used by Columbus as privileged channels of access to the King and Queen: the Prince's nurse, Juana de Torres y Ávila, was a confidante to whom Columbus—to judge from his surviving letter to her, written in 1500—could unburden almost as freely as to Deza. In his letter he treats her as privy to his dealings with the Queen. Her brother, Antonio de Torres, though not in a position of comparable influence, was Columbus's messenger to the monarchs in 1494, sent back from Hispaniola, after accompanying Columbus's second crossing of the Atlantic, to present Columbus's defence against his detractors. Juan de Cabrero, who was transferred from the household of the Prince to the post of chamberlain to the King, was credited in early sixteenth-century traditions with a vital role in procuring royal sanction for Columbus's plans. What exactly he is supposed to have done is not made clear but he had an unrivalled opportunity to exercise influence in his daily closetings with the King, at his *toilette*, and during their after-lunch conversations, when— according to another old member of the Prince's household who followed Columbus to the Indies—'the King would have a little seat brought for him on which he sat and discussed matters with the King, cordially, as with a man whom he loved'. The head of the Prince's court, Gutierre de Cárdenas, son-in-law of the Admiral of Castile and holder of some of the highest offices of court and state, is regarded by early tradition as a friend and supporter of Columbus; though this reputation cannot be substantiated from surviving documents, it seems credible in view of the role of the princely court as the hive of the Columbus faction. Cárdenas's business interests in Canarian dyestuffs seem to have brought him into close contact, moreover, with some of the financiers in Quintanilla's circle.[32]

A third group whose support Columbus won was connected with the seaport of Palos, where the shipping, manpower, and logistical support for his enterprise were to be concentrated. The key figure in the growth of this group may have been the Franciscan Fray Antonio de Marchena, the only court astronomer to credit any of Columbus's geographical speculations. Columbus later paid tribute to the singularity of Marchena's help in a letter to the monarchs: 'Your Highnesses already know how I wandered for seven years in their court, entreating them to this design . . . and never in all that time was there a pilot or mariner or philosopher or other expert but said my proposal was false, and that I had help from none save Fray Antonio de Marchena.' The monarchs even suggested that the friar should follow Columbus across the Atlantic 'for he is a good astronomer and it always seemed to us that he agreed with your opinion'.[33] Marchena was head of the Franciscan province of Andalusia and at one time guardian of the friary of La Rábida, by Palos, overlooking the point where the Guadalquivir flows from Seville into the Atlantic. It was a severe and sequestered community, but Marchena was not its only link with the world of the court. Another of its guardians, Fray Juan Pérez, was one of the Queen's confessors. Columbus was there in the summer of 1491: the legend that makes him call at the friary by chance for the first time six years earlier, begging sustenance in his poverty for his fainting son, probably mixes the confused testimony of forgetful witnesses with the smack of romance.[34] The visit of 1491, however, was of decisive importance in advancing the explorer's plans. It was the occasion of consultations with a local physician and astronomer, García Fernández, with the shipowner who was to provide the means of Columbus's voyage, Martín Alonso Pinzón, and perhaps with a seaman, usually called Pedro Vasques, who was a source of tales of fleeting landfalls deep in the Atlantic, and who, as the pilot of the discoverers of Flores and Corvo, could be presumed to have more useful yarns than the average ancient mariner.[35] Fray Juan then communicated with the court, perhaps journeying in person to make new remonstrations on behalf of the project. The effects of his intervention were significant; the Queen sent money to clothe Columbus suitably for a new audience and granted him permission to hire a mule to ride to her presence. This was a considerable privilege, conceded only because Columbus's straitened circumstances had impaired his health, for the monarchs were fostering a war effort and enforcing austerity; the use of mules was restricted.

There were other figures whose support for Columbus can be perceived only dimly. The Primate of Castile, Archbishop Pedro González de Mendoza, is credited with support for Columbus in all the early secondary accounts, but it is not clear why, or with what justification. The royal confessor, Fray Hernando de Talavera, who actually headed the committee of experts which had rejected Columbus's plans, was regarded by Peter Martyr of Anghiera—the Italian humanist employed by Ferdinand and Isabella—as instrumental in the explorer's success and by Las Casas as 'helpful'. The only help he is known to have given was in handling some of the money raised for the first voyage for safe keeping, passing it on to Columbus.[36] Perhaps in self-flattery, Columbus liked to include the Queen herself among his longstanding friends at court. 'Everyone else was disbelieving, but to the Queen, my lady, God gave the spirit of understanding . . . and great strength, and made her heiress of all, like a very dear and well loved daughter.' If one were to add up all the solitary individuals whom Columbus is supposed to have identified as progenitors of his success, there would be a sizeable cohort: Diego Deza; the unnamed 'two friars' (one of whom most commentators identify as Fray Juan Pérez); Fray Antonio de Marchena (cited by name); Juan de Cabrero; the treasurer of the Crown of Aragon (to whose case we shall turn in a moment); and the Queen herself. Not all these can have been uniquely effective, and we are clearly in the presence of a rhetorical device. Columbus often used his claim to be particularly indebted to the Queen in order to cast the King's efforts in a contrasting and unsatisfactory light; yet objective evidence shows the King to have been just as generous to the Columbus family after the Queen's death as ever the united couple were in her lifetime.[37]

Columbus's unctuous assertions about the favour of the Queen have influenced historical tradition profoundly, but have to be treated as unverified. He was a master of the coquettish language the Queen liked to hear. 'The keys of my desires I gave to you in Barcelona. If you try a taste of my good will, you will find its scent and savour have only increased since then . . . I dedicated myself to your Highness in Barcelona without holding back any part of me, and as it was with my spirit, so it was with my honour and estate.' The tone seems to match the flirtatious atmosphere of a court where—as in that of Elizabeth I of England—the Queen's very reputation for chastity licensed verbal gallantry. Columbus appealed to the Queen in prose in terms very similar to those chosen, for instance, by Juan Álvarez Gato in verse:

> My soul is fasting. I appeal
> To you for aid. I'm near to dying.
> All the world knows how my weal
> Is one that you alone can heal.

Or again:

> You are paramount in beauty
> Whereas I in love am first.
> Than your estate is none more great,
> While I with greatest grief am cursed.

Columbus seems, therefore, to have known how to get attention from Isabella, but we cannot trust his or his eulogists' accounts of how she responded.[38]

One further group of Columbus's partisans can be distinguished clearly, centred on the person of the treasury official of the Crown of Aragon, Luis de Santángel, and probably his colleague, Gabriel Sánchez.[39] Santángel had acquired some financial responsibilities within the Kingdom of Castile, especially in managing the affairs of the local militias and in connection with the sale of indulgences. Close collaborators in these duties were Alonso de Quintanilla and Francesco Pinelli. Santángel's importance for Columbus seems to have been in his role as a financial fixer, bringing together the sources of investment and means of raising cash that finally made the proposed voyage possible. The sum that needed to be found was not outrageous: two million maravedis—perhaps the annual income of a middling provincial aristocrat; but the enterprise was risky, the pundits were derisive, and in time of war cash was short. The money would have to be pieced together, and any conspicuously large sums carefully indemnified.

Columbus always complained that the monarchs 'would not put up more than one million'. But this was highly misleading. The total 'public sector' contribution amounted to 1,140,000 maravedis, including 140,000 earmarked for Columbus's wages. The entire sum was advanced by Santángel and Pinelli against expectations from the sale of indulgences and, in fact, it was all recovered in due course from the proceeds of sales in a poor diocese of Extremadura. Part of the balance was provided in kind by Martín Alonso Pinzón and his collaborators in Palos: the town owed the Crown the use of two caravels by way of a fine. The rest, including or equalling the 500,000 maravedis of which Columbus himself was the nominal stakeholder, was supplied by Quintanilla's syndicate. At least some of it came from Gianotto Berardi, whose will, dated three years later, recorded an outstanding debt from Columbus of 180,000 maravedis: Berardi, however, handled a great deal of Columbus's business during his absence, and it would be rash to assign that particular sum specifically to the expenses of the first voyage. Thus the Queen did not have to 'pawn her jewels' to pay for Columbus's crossing: in view of the burden shouldered by the impoverished penitents of Extremadura, that legend seems a particularly impious fiction. Nor did the Crown have to find any ready money. By January 1492 the financial obstacles to Columbus's departure—which, in the final analysis, were the only obstacles that mattered—had disappeared.[40]

By or during 1492 the scope of the transatlantic project narrowed. The Antipodes and unknown islands ceased to be advocated as objectives of the quest. The short route to Asia became the only possibility to qualify for further mention. This suggests that the concentration on an Asiatic objective and the long-delayed success of the search for patronage are connected. It is safe to assume that virtually nobody at the court of Ferdinand and Isabella was anxious in 1492 for the discovery of more Atlantic islands. The last isles of the Canaries, Tenerife and La Palma, were proving obstinately hard to conquer. The difficulties experienced by the Portuguese in the Cape Verde group and the remoter Azores suggested that further discoveries would be hard to colonize, unless exceptionally attractive: in any case, the newly won territory of the Kingdom of Granada had to be sewn with Castilian settlers as a first priority. The discovery of the Antipodes, even for those who believed that such a destination existed, could not be represented as offering any specific prospect of profit. An unknown land represented, by definition, an unknown risk. Only the Orient, with its connotations of gold and spices, was potent enough to engage investors' eyes. Ferdinand and Isabella, smarting under the costs of the Granada war, envious at the profits got by the Portuguese in Guinea, were more likely to back a venture which promised—however insecurely—lavish returns and a march stolen on their rivals than to take an interest in avowedly unprofitable exploits. In a document drawn up for himself in 1492—the petition on which the terms of his commission from the monarchs was based—Columbus left his destination vague. In his communications with the monarchs, he omitted no opportunity to stress his commitment to the search for Asia. The simplest explanation of the narrowing of his plans is that it was done to please his patrons.

The growth of the Columbus 'lobby' has been traced and its components analysed, but the problem of how it was created remains. Evidently it was a slow process, of which Columbus was inclined, at intervals, to despair. In 1488 he returned for a time to Portugal, to offer his project there; in about 1489 he sent his brother Bartolomé to England and France for the same purpose, with equally little success. On at least one other occasion, if his later memories can be trusted, he was resolved to abandon Castile, but was dissuaded by Diego Deza.

Slow grinding, however, is effective. Columbus's supporters formed a spreading network which grew by personal contacts, individual introductions, and the operations of friendship and common interest. The kernels of his support—among the Genoese and Florentine merchants of Seville, the friars of La Rábida and the court, the financiers of the conquest of the Canary Islands, the household of the Prince Don Juan, the treasury of the Crown of Aragon—overlapped and formed an ever-strengthening web. Moral support gradually merged into political influence and eventually produced financial backing.

It is impossible, however, to imagine Columbus merely as a plaything manipulated by rich investors and powerful politicians. His extraordinary personal gifts—imaginative, persuasive, even perhaps charismatic—have to be taken into account when the problem of how he launched his enterprise is addressed. Anyone who reads Columbus's writings can share the impression he made on his contemporaries: he emerges as a man blessed with gifts of natural rhetoric and tireless eloquence—or, at least, indefatigable loquacity. His errors, fallacies, and articles of faith were all expounded with undaunted conviction. He was possessed of a power of self-assertion which no mockery could challenge and no failure quench. His appearance—tall of stature, sanguine of colouring, and pale-eyed—made him conspicuous; he became a familiar figure in the court of the Catholic monarchs and in Seville and Cordova, where he lived when not trailing, in irrepressible optimism, behind the royal camp. Columbus was the sort of man with whom familiarity bred respect—respect for his conviction, respect for his experience, respect, if for nothing else, for his persistence. The years of his quest for patronage in Castile were not, as he later claimed, a lone struggle against the odds, but a gradual gathering of support in an increasingly favourable context. Still, they were a story of personal triumph, gained, in part, by individual effort and merit.

In the Columbus romance, an irresistible and incredible story ends this period of Columbus's life. On his return to court yet another committee of experts was convoked to listen again to his pleas, in circumstances of high drama, in the royal camp during the last siege of Granada. Against this spectacular backdrop, the theatrical encounter is played out. Another peripeteia dashes Columbus's fortunes: the experts have declared against him once again. On the second day of the New Year 1492, Ferdinand and Isabella ride into Granada as conquerors. Only Columbus, of all those privileged to assist, cannot share in the rejoicing. He turns his back on the triumph, and rides disconsolately towards La Rábida in the knowledge that his suit has finally failed. Then the most romantic touch of all: after a day on the road he is overtaken by a royal messenger who demands his immediate return to the monarchs' camp. A change of heart at court has come suddenly, and against expectation, like all the best miracles. Columbus makes the first leg of his Atlantic voyage by mule to Granada.

4

'The Conquest of What
Appears Impossible'

THE FIRST ATLANTIC CROSSING,
AUGUST 1492–MARCH 1493

ABOUT seventeen years later, long after Columbus had made his last voyage over the sea of death, a young Spaniard, noble both by nature and blood, who had settled successfully in the New World discovered by Columbus, sat listening to the heated preaching of a Dominican in the city of Santo Domingo, founded by Columbus to be the effective capital of the Spanish New World of the day. Bartolomé de Las Casas— according to his own recollection of the event— experienced as he listened a sudden revelation. He found himself sharing the friars' outrage at the corrupt lives of the colonists and their ruthless exploitation of the wretched natives. He became convinced that God had ordained the discovery only so that the Indians should be able to hear His word preached. Las Casas devoted the rest of his energetic and ineffective life to his conviction, becoming a lobbyist at court for Indians' rights, and at intervals a rather unsuccessful pastor in the field in the New World, struggling to liberate and hallow the lives of Indians already under Spanish rule and to promote the evangelization of others.

Among the many books he wrote in pursuit of his aims was a compendious *History of the Indies*, which he compiled—it is true—*à parti pris*, but with laborious fidelity to the sources. Among those sources, the most assiduously pillaged were writings of Columbus's own. Columbus had for Las Casas a special place in the sacred story as the instrument whereby the divulgation of the gospel was made possible in the New World. Columbus's own sense that he was the executor of a divine mission appealed to Las Casas and won his sympathy. The Dominican neophyte therefore read, marked, noted, and in many cases copied or abstracted as much of the discoverer's writings as he could find. Except for a few fragments preserved in other sources, Columbus's copious relations of his voyages made on board ship for the information of his royal sponsors now survive, thanks only to the summaries, transcripts, and paraphrases made by Bartolomé de Las Casas. To his piety we owe most of what we know of Columbus's navigations.

We are particularly fortunate in relation to the first crossing because Las Casas preserved for his own *History*, the gist of an account called the *Libro de la primera navegación*, or *Book of the First Navigation*, which was either a copy of Columbus's shipboard work, or an edited version of it, prepared perhaps with an eye to publication. It never was published and the manuscripts have vanished or perished, but Las Casas's abridgement, which he called *El primer viaje* or *The First Voyage*, transmits more of Columbus's account than survives for any subsequent voyage. Four distinct voices can be detected in it. There are parenthetical or marginal comments inserted by Las Casas, which are occasionally hard to distinguish from the rest of the text, but which are usually marked by their sententious tone: the editor marvels, for instance, at his hero's insensitivity to the evils of slavery, or praises his unshakeable faith in God. There are passages of paraphrase in which Las Casas narrates the events without disclosing the relationship of his own text to the original: sailing directions tend to be treated in this summary fashion, exacerbating the difficulties of reconstructing Columbus's routes. Other passages are in reported speech, often with discernible echoes of Columbus's own language. Finally, there is fairly prolific direct quotation, almost all of which is about the Indians; this is undisguisedly selected to reflect the editor's priorities rather than the author's and confirms—on the whole— Las Casas's image of the Indians as uncorrupted, peaceful inhabitants of a world of sylvan innocence, naturally good and unconsciously dependent, in their defenceless nakedness, on the loving mercy of God. Despite these textual problems, and the inevitable corruptions introduced into a text avowedly based on a copy, *El primer viaje* remains a wonderfully vivid and exciting document, which is supported, and sometimes supplemented by passages included in the *History of the Indies* or in other early works based on materials identical with or similar to those at Las Casas's disposal. On its own, it is enough to establish Columbus as a man of extraordinary character and exceptional gifts. His reports to Ferdinand and Isabella are unique in the annals of the sea; no master ever compiled so detailed a log; no commander of the day ever wrote such copious reports; no navigator of that era—except perhaps Columbus's future rival, Amerigo Vespucci—displayed such talent for observation, such sensitivity to the elements, such appreciation of nature. No sailor ever revealed so much of himself in a yarn. Columbus was a 'poor foreigner' who never perfectly mastered written Spanish, as Las Casas complained; he was an autodidact who was never taught the craft of writing. But he combined natural rhetoric with a responsive spirit and he had a compelling tale to tell.

The First Voyage opens with a 'Prologue', addressed to Ferdinand and Isabella, which purports to record his emotions at and just before his departure.[1] In reality it is evidently bodged together from two or three documents, widely differing in time and purpose. It does, however, capture faithfully Columbus's own understanding of the purposes of the voyage. At a superficial level, it proclaims the objective of reaching 'the lands of India and of a prince called Great Khan' not 'overland to the east, which is the customary route, but by way of the west, by which route to this day—as far as we know for certain—no one has ever gone before'. At a deeper level, the enterprise is justified in the religious terms calculated to appeal to the monarchs, as part of an evangelical and crusading mission inherited from the past. But what marks it as authentically the work of Columbus, and a reflection of his own most profound concerns, is a passage in which he takes the opportunity to record the terms of the commission in which the monarchs had promised him, in return for success, the noble status he craved:

and to that end your Highnesses granted me great favours and ennobled me, so that from thenceforth I should be entitled to call myself Don and should be High Admiral of the Ocean Sea and Viceroy and Governor in perpetuity, of all the islands and mainland I might discover and gain, or that might thereafter be discovered and gained, in the Ocean Sea, and that my elder son should succeed me and his heirs thenceforth, from generation to generation, for ever and ever.

In today's language, Columbus was 'writing to confirm the terms of his contract'. He was more concerned to record his titles of nobility than the substantial economic concessions he was granted, conditional on success. For the terms of his commission survive in a copy he kept himself and can be compared with his own summary. While Columbus arrogated to himself the immediate and unconditional use of the promised titles, it was the monarchs' intention that his aspirations to nobility should be satisfied only if his voyage proved profitable. The office of admiral was conferred with jurisdiction over the Ocean Sea on the same terms enjoyed by the hereditary admirals of Castile's home waters—an extraordinary privilege to be shared by a weaver's son with the head of one of the noblest families in the kingdom; yet strictly, amazing as it seems, Columbus appears to have been justified in making explicit the assumption that this was to be a hereditary dignity. The use in the commission of the title of viceroy, unprecedented in Castilian usage, tends to confirm the impression that the royal chancery adopted a draft submitted by Columbus without detailed editing and perhaps without much thought or scrutiny. This alone helps to explain the insecurity Columbus evinced in writing his own summary record in the prologue to *The First Voyage*.

Moreover, he may have been responding to the chancery's first attempt to claw back what had been granted him when his contract was recast in the form of a letter of privilege—a royal grant and therefore revocable—of 30 April 1492. Only after his return from his voyage of discovery, in May 1493, did a royal letter make explicit the extension of his functions from the islands to the open sea within the limits of his admiralty, while, finally, it was not until 1497 that the concession of a *mayorazgo*, or right to make his whole estate heritable by a designated beneficiary, had the effect of confirming the hereditary nature of his offices and protecting them from division or revocation.[2]

Still, the economic and jurisdictional rights which he did not bother to record in the prologue were potentially immense. The grants were of two types: pecuniary perquisites on the one hand and, on the other, political privileges of a distinctly feudal flavour. On the economic side, Columbus was to receive one-tenth of the profits of his admiralty over and above the other dues consequent upon the office of admiral— though in practice he never received all these. Where his formal powers of feudal eminence are concerned, the source of Columbus's power lay in the conjunction of the offices of admiral, viceroy, and governor and their being made inseparable and hereditary. The effect, one could suggest, was to turn the Ocean Sea and all its lands into a potential feudal seigneury only a short way removed from a principality. Columbus was to have first, in his own sphere, all the rights of jurisdiction of the admirals of Castile, consisting of dispensing the highest form of justice and imposing the penalty of death—though he was to be penalized for the alleged arbitrary exercise of the last right. He also possessed the right of pardon and could judge cases arising in Castile in connection with the Ocean trade. The nomination of subordinate officers and ministers of justice Columbus did not enjoy in its entirety, but could only present a short list to the monarchs. It seems doubtful, however, whether this modification applied in practice and direct appointments were made by both Columbus and the Crown. The organization of fleets Columbus was to share with royal nominees, but, as events were to prove, a great degree of liberty of navigation was inevitable in the early years of the discovery. As viceroy and governor, Columbus would be able to enjoy similar rights of jurisdiction and patronage and could, in theory, command the obedience due to the monarchs. His dignity as viceroy, being hereditary and inseparable from his other charges, was superior to that of Aragonese viceroys, on whose office his was presumably modelled.

It seems that Columbus gave rather more thought to the implications of these concessions than did the monarchs. Later, as the enormous extent of his discoveries was revealed, Ferdinand and Isabella were obliged to disregard or supplant many of his offices and powers. Theoretically, the only means open to them to restrain him was the judicial investigation of a governor's conduct, the *residencia* or *pesquisa*, to which he was submitted twice during his term. He was the only official of the Spanish Crown up to that time, apart from Alonso de Lugo in Tenerife, to undergo such a process while in office. The same means were used to keep his son and successor, Diego Colón, in check. The fact that Columbus was never able to rule his domains in the surly independence attained by some past and future conquistadores was partly a consequence of the fact that he was to head an underdeveloped and unruly colony, where his local opponents were adept in exploiting the opportunities of appeal to the Crown, or to interested royal officials, in thwarting Columbus's orders. He had every temptation to exceed his powers, and few chances to enjoy them. He was the victim, moreover, of his own success in imposing on the monarchs in 1492 terms so generous as to make them apprehensive and alert to the need to curtail his powers. But most of all, as we shall see, he was limited by his own incompetence, and, rather than consolidate his power, he preferred to flee his responsibilities by undertaking new explorations or withdrawing to Castile. Gradually, therefore, and increasingly, the monarchs were able to interpret the terms of their grants in a way most unfavourable to Columbus, depriving him of his privileges and breaking the exclusivity of his titles and offices. After an arduous litigation, the Colón family finally yielded their claims to Columbus's privileges, long after their substance had been lost, in 1556. In more dramatic terms, the episode can be seen as the triumph of a centralizing monarchy over a feudal tendency at the periphery of its empire.

These ramifications, however, were unforeseeable when Columbus left court for the coast with his commission under his belt, and raised sail, out of Palos, over the bar of Saltes, on 3 August 1492. The scale of the expedition reflected the low expectations most people had of it: three small vessels, a few more crew than the eighty-eight enumerated in our most reliable tally. The monarchs had ordered the town of Palos to provide two caravels as quittance of a fine which the municipality owed the royal treasury. The first of these was the square-rigged *Pinta*, so-called perhaps after the brothers Pinzón, and the other the *Niña*, owned by Juan Niño, who was also to sail in her on the ocean crossing. The detailed appearance of the vessels is unknown and all 'reconstructions' fanciful.[3] The *Niña* was a fast, trim vessel of moderate size, rigged with triangular sails, and the *Pinta* of about the same bulk but slightly slower. The largest ship, though not by much, was the ill-destined flagship *Santa María*, with her round hull and ponderous gait, and the monograms of Ferdinand and Isabella on her mainsail. She was the only one of the ships known to us by her proper name. Her nickname, *Gallega*, probably reflects not her ownership but her build, in a Galician port. The crews included a contingent of Basques but were recruited mainly in Palos and Seville, probably by Martín Pinzón, who captained the *Pinta* and who, as the head of the Pinzón clan, had sufficient prestige in Palos to overcome the apprehension of potential recruits for a voyage into the unknown. The task of recruitment had been eased by a royal pardon to any condemned man who shipped on the voyage but in practice little recourse was had to any such dubious source of man-power. Pinzón's role was vital and it can be inferred that, whatever the acrimony that arose between him and Columbus in the course of the voyage, they started on terms of mutual confidence. The effect of his responsibility for recruitment was, on the other hand, to crew all the ships with his creatures so that when he and Columbus fell out the commander was fearful, exposed and almost isolated.

No soldiers or settlers accompanied the fleet, for the expedition was consciously one of exploration. Before departing, Columbus loaded a great store of truck, which he hoped to replace with samples of spices and gold: the hawks' bells, beads, and glass he carried were more attuned, however, to the experience of the Portuguese in West Africa than to the needs of the sophisticated Oriental markets in which he was hoping to trade. Though he expected to revictual in the Canary Islands, he took on the salt fish and bacon, biscuit and flour, wine, water, and olive oil that sustained all Mediterranean seamen.

The course via the Canaries was the master-stroke of the voyage. Before Columbus could discover America, he needed to make what was, in a sense, a far more transcendent discovery: that of a route across the Atlantic (see Map 3). A point of departure further north would have made it impossible for him to find a following wind. By choosing the Canaries he created the chance of success. It is probably fair to say that—taking the history of exploration as a whole—most exploratory voyages have been made against the prevailing wind, except in monsoonal climes, because it is as important to the explorer to find a route home as it is to make a new destination. It is precisely their departure from this rule that makes some of the most conspicuous voyages in history—those of the Vikings around the extreme North Atlantic, of the Spaniards of the sixteenth century across the Pacific, and of Columbus to the New World—seem so daring and heroic. Apart from that of Columbus, the attempts at further Atlantic exploration in the fifteenth century that we know about set out from the Azores or Bristol, into the belt of westerlies; and most, in consequence, failed. Columbus's course, by contrast, was almost perfectly judged, in both directions. He left from a point near the twenty-eighth parallel, where he could be almost sure of having the north-east trades behind him; on his return, he began by striking north, looking for the westerlies that would bring him home, and finding them promptly. Except for his failure to exploit the Gulf Stream, which was not discovered by European navigators until 1513, he was, by the time of his second voyage in 1493 (when he improved on his original course by crossing the ocean slightly more obliquely) almost exactly foreshadowing the standard optimum galleon route of the remainder of the age of sail.

This uncanny good fortune has suggested to some historians that he must have had secret foreknowledge of the route, imparted by an 'unknown pilot' or culled on some unknown previous voyage of his own. Recourse to such fanciful explanations is not necessary, though some other reasons commonly adduced are equally unsound. He did not, for instance, choose the Canaries because he expected to find Cipangu on the same latitude: it is clear from his log that he believed Cipangu lay south of his chosen course.[4] Nor did he choose the islands simply because they had a suitable harbour in a relatively westerly position. The Azores and Cape Verde Islands lie much further west and the Azores, because they are well to the north, promise a shorter run along the diminishing curve of the earth than the relatively southerly Canaries. Two possible explanations of Columbus's route are sufficient, each in itself, and one can choose between them, if one so wishes, according to philosophical inclination. Individualists will appeal to Columbus's sagacity, pointing out that he was well schooled in the ways of the Atlantic and had plenty of opportunity in the late 1470s and early 1480s to observe the rudiments of the wind system. Determinists will counter that Columbus had no choice; once he had thrown in his lot with Castile he had to sail from a Castilian port, and the Canaries were the only Castilian possession in the Atlantic.

The passage to the Canaries was rapid and uneventful. For the major part of the crossing, Columbus had the *Niña* converted to square rigging to catch the following wind—another indication that he knew the sort of winds to expect. His course through the islands seemed dilatory, but it was important to get the vessels shipshape for what promised to be the longest voyage on the open sea ever recorded. In addition to regular maintenance and the conversion of the *Niña*'s rig, it was necessary to replace the *Pinta*'s rudder. More supplies had to be loaded, including no doubt some of the Gomeran cheeses that were particularly suitable for ships' victuals. And the expedition had to endure a tense wait for a favourable wind. When the easterlies sprang up, on Thursday, 6 September 1492, the explorers quit San Sebastián de la Gomera amid a flurry of square sails and, leaving the isle of Hierro to port, took their leave of the known world.[5]

Columbus set his course due west. His intention—albeit not sustained in the event—was to maintain the same course until he struck land. This plan presented no serious problem of navigation. To sail a straight course demanded only the sort of know-how that derives from experience: it was a matter of observing and making allowance for the effect of wind and current, with the aid of the compass that all ships of the time carried. Still, Columbus's methods of navigation presented some curious features. He is commonly characterized as merely a dead-reckoning navigator: one, that is, who observes his course by the compass, records the time travelled in any direction, and estimates the speed of the ship to arrive at the distance traversed.[6] Even this rather basic method demands considerable skill if it is to yield a reliable record of a course across open sea. Apart from celestial observation, the only method of recording time aboard ship was by means of the sand-filled hour-glasses turned at half-hourly intervals by ships' boys, whose negligence or excessive zeal could ruin the calculations. To estimate speed with even the crudest semblance of accuracy was a highly esoteric art on the open sea, with no fixed landmarks—and even, for most of the time, no flotsam—to judge by. It is remarkable that in the early centuries of the age of sail dead reckoning could yield such accurate results, or that sailors could ever retrieve a known course or recover a previously visited destination.

Even without the use of navigational instruments or charts, dead reckoning could be enhanced or replaced by what maritime historians call 'primitive celestial navigation'.[7] In the Northern hemisphere, with the Pole Star for a guide, this is a relatively easy craft (though it has been irretrievably lost today) which was cultivated over generations, and improved with practice, by late medieval mariners in Latin Christendom. Its potential has been demonstrated in our own times, in the Southern hemisphere where the heavens are much harder to read, by Polynesian seafarers, who can record and retrace a course across thousands of miles of open ocean, with only their mental star-map, and the subtle, sensibly discerned features of the wind and sea to guide them. The people of the Marshall Islands keep marine charts made of reeds which relate islands to each other, to wave patterns, and to navigable routes; indigenous navigators of the Pacific, however, can sail without such aids, or navigational instruments. In a conscious experiment, the navigator Piailug, from the Micronesian atoll of Satawal, sailed a Hawiian double canoe from Hawaii, over 3,500 miles of open sea, to Tahiti, between islands and through seas unknown to him, without error. The only information he had at his disposal were star bearings.[8]

Columbus's task in keeping a straight course could easily have been accomplished by the unaided eye of a practised celestial navigator: it would have been necessary only to keep the sun, by day, and the Pole Star, by night, at a constant angle of elevation. The evidence of late medieval marine charts makes it clear that navigators of that era were well able to judge relative latitudes with the naked eye to within a serviceable margin of error. Columbus himself claimed to have done so during his return voyage, on 3 February 1493.[9] Whether, however, from the insufficiency of his technique in primitive methods, or from scientific curiosity, Columbus resolved to go beyond dead reckoning and primitive celestial navigation in three ways: by the use of a chart; by trying to take exact astronomical readings of latitude; and by verifying latitude by timing the length of a solar day.

A chart might have been supposed to have been of little use in literally uncharted waters. Yet Columbus carried one, speculatively constructed on the basis, it can be inferred, of small-world theory and of the existence of Cipangu; if it resembled a map Columbus later referred to in correspondence with an English informant who reported to him the discoveries of John Cabot, it probably also showed Antillia—'the Isle of Seven Cities'—a mythical refuge first located in the mid-ocean in a map of 1424. Las Casas was convinced that Columbus's chart had been made by Toscanelli: if so, it would have been constructed on a grid of lines of longitude and latitude, purporting to convey a good notion of distance, in contrast to the common mariner's chart of the late Middle Ages, which showed only courses in the form of rhumb lines leading from wind-roses at strategic points; the effect was to create delicate, intersecting, web-like patterns, but, despite its aesthetic attractions, this type of chart was almost oblivious of considerations of distance. Columbus's chart inspired in him the most extraordinary misplaced confidence: at intervals during the voyage he consulted over it with Martín Pinzón, contemplating—and once, at a crucial juncture, actually effecting—a change of course on the strength of it, and apparently making from it judgements about their proximity to land. This last point should not be over-emphasized however: Columbus may have been misusing a conventional sea-chart or simply using the map to legitimate his own calculations, based on his small-world theory and his sense of how far he had gone.[10]

The Admiral's use of navigational instruments is captured in a passage, presumably derived from the Book of the First Navigation but omitted from The First Voyage and preserved only at second hand, relating to 24 September 1492, when, after a series of phoney landfalls, Columbus's fears of mutiny were at their height. According to Las Casas's graphic account, conspirators were murmuring that 'it was great madness and self-inflicted manslaughter to risk their lives to further the mad schemes of a foreigner who was ready to die in the hope of making a great lord of himself'. Some of them argued 'that the best thing of all would be to throw him overboard one night and put it about that he had fallen while trying to take a reading of the Pole Star with his quadrant or astrolabe'.[11]

The story these mutineers are alleged to have concocted has an irresistible grim humour. It brilliantly evokes the figure of the outlandish boffin, practising in ungainly isolation his new-fangled techniques, while struggling on a rolling deck with an unmanageable gadget. This image, which so annoyed the plotters, conveys a good deal of truth.

Columbus, despite his pride in his navigational instruments, was never able to use them properly, defeated by the pitch and roll of the ship. The readings he claimed to make with his astrolabe were in reality based on a less glamorous method. He calculated the length of the day, in terms of hours of sunlight, and read his latitude off the table he had copied in Pierre d'Ailly's *Imago Mundi*.[12] The mistakes he made exactly correspond to misprints in the table.[13] He was, understandably, evasive about his recourse to this crib, but explicit about his means of reckoning time. The ships' boys could not be trusted with the sand-clock, but the celestial bodies kept reliable time. Every twenty-four hours, the Guards in the constellation of the Little Bear describe a complete revolution around the Pole Star. The naked eye is sufficient to tell roughly how far round their course they have got at the time of a particular observation. As a memory-aid and standard of measurement, medieval navigators divided the circle into eight equal divisions, normally named after the parts of the body or the points of the compass. Thus the 'north-east' division, at 45 degrees, was called 'the right shoulder' or 'above the right arm', the 'east' division 'the right arm'; the 'south-east' was 'below the right arm', and so on. By observing the movement of the Guards across these divisions, Columbus could calculate the length of the night and so, by subtraction from twenty-four, the hours of daylight. On 30 September, for instance, Columbus tracked the Guards across three of the divisions, establishing a figure of nine hours for the duration of the night and fifteen for the hours of daylight.[14] His obsession with the hours of sun is intelligible only as part of his concern—a mapmaker's concern, stimulated perhaps by his promise to chart his discoveries for the monarchs—with the assessment of latitude.

He also kept a wary watch on the Pole Star, and stumbled as a result on one of the most important cosmographical discoveries of his day. Preserved in *The First Voyage*, Columbus made a series of recorded observations of magnetic variation: the difference, that is, between the direction shown by the compass needle (magnetic north) and the 'true north' indicated by the position of the Pole Star. In the Eastern hemisphere, variation to the east was a familiar phenomenon. Columbus's are the first known recorded observations of variation to the west. On 13 September he recorded slight variation in both directions, perhaps because he was passing between zones. From 17 September onwards he noted strong and increasing variation to the west. The question arises, did he recognize this phenomenon for what it was? The same question has been asked in a larger context of his discovery of

America, and it is useful to see Columbus's mind grappling with the problems—conceptual and classificatory—of fitting a new observation into the framework of existing knowledge. His first response seems to have been entirely practical: he had to guarantee his crews' confidence in the reliability of the compasses and therefore sought to minimize the problem by taking readings at the most favourable time of day, when the Pole Star was apparently in its most westerly position. This, together with his sustained interest in making readings and noting the results, suggests that he at least considered the possibility that the apparent fluctuations were genuine. By 30 September he seems to have resolved the problem in his own mind by explaining it as the apparent result of the instability of the North Star: 'the North Star moves, like the other stars, while the compass needle always points in the same direction.' To judge from the language of this sentence, it was an explanation in which he believed, not a mere deception concocted for the benefit of his crew.[15]

So much for the stars. What was happening on the surface of the ocean during the crossing? Columbus's account is dominated by four recurrent themes: the phoney landfalls, which undermined the men's morale; the fears, as the winds carried them swiftly westward, that they would never find a wind to take them home; the increasing tension between Columbus and Pinzón and between commander and crew; and Columbus's own barely perceptible but genuine doubts, which afflicted him increasingly as the expedition spent longer and longer out of sight of land. They rapidly negotiated the dangers of a Portuguese squadron sent to intercept them and the mysteries of the Sargasso Sea, of which they could have had preliminary notice to allay their fears without diminishing their amazement; but the insidious element of uncertainty about their destination and about the advisability of their voyage made the fair passage a time of torment. In mid-September, Columbus's search for signs of land, in the form of the swirl and fall of songbirds, seems to have nourished heightened perceptions and to have blended with echoes of Noah's Ark:

Thursday, 20 September: He set course this day west by north and at half sail because conflicting winds succeeded the calm. They would make seven or eight leagues. There came to the flagship two herons and later another, which was a sign of nearness of land. By hand they took in a bird which was a river-bird, not of the sea, though its feet were like a gull's. There came to the ship at dawn two or three landbirds singing and later before daybreak they departed.

In Columbus's implicit self-comparison with an Old Testament patriarch it may not be fanciful to detect the first glimmerings of his

growing conviction—first made explicit on his return voyage in 1493—
that he had a sort of personal covenant with God. On 23 September he
reported a 'high sea, the like of which was never seen before except in
the time of the Jews when they fled from Egypt behind Moses'.[16]

The apprehension of danger seems only to have heightened
Columbus's sensibilities. His most rhapsodical response to his environment occurred in a passage summarized by Las Casas under 16
September: 'the scent of the morning gave real pleasure and the only
thing wanting was to hear the nightingale sing, he says.' Readers of
Columbus's sensitive praises for the perfections of the Atlantic world
have sometimes wondered whether they should be traced to the
influence on Columbus of Franciscan values, with their strong rever-
ence for creation, or to Renaissance aesthetics, which are associated
with interest in the realistic depiction of the natural environment. All too
often, however, Columbus's attempts to evoke the beauty of nature can
seem crass—as here, where the nightingale is associated with the
morning—or nebulous. His descriptions, in particular, of the landscape
and flora of the Caribbean remind one of nothing so much as of Milton's
paradise. It should be remembered that Columbus was writing promotional literature, designed to boost the image of the Ocean and attract
further investment and royal favour to his enterprise. He had an interest
in stressing a salubrious environment, because it would help to make his
route exploitable. This is not to say that he was personally unconvinced
of the veracity of his descriptions. His often-voiced conviction, for
example, that the air and climate perceptibly improved a hundred
leagues west of the Azores was sustained so consistently as to defy
scepticism, though it must have owed more to his imagination than to
any measurable effect.[17]

Columbus soon half-revealed to himself his own doubts of the
distance to the Indies, for from 10 September he began to falsify the
ship's log, undercutting the number of miles they had covered in the
estimates he retailed to his men. He revelled in his role of lone
manipulator with evident relish and feigned unease. He positively
enjoyed the cybernetics of deception: his own proud memory of the crew
he duped, as a young man, into mistaking Tunis for Marseilles[18] is
recalled by the story of the doctored log. These episodes remind the
reader of how much we depend, for our knowledge of what happened on
the voyage, on sources directly inspired by Columbus himself. The
images of isolation and vulnerability are projections of his own self-
perception. The atmosphere of conspiracy could be the product of a

paranoid imagination. The evidence of the deepening rift with Martín Pinzón—which is rigorously excluded from *The First Voyage* and has to be pieced together from other sources, including independent legal testimony of much later date—could have sprung in the first place from his own mistrustful nature and a loner's dislike of collaboration.[19]

In fact, since Columbus's approximations of distance always tended to be overestimates, the falsified log was more accurate than the private one he kept for himself. His weakness for wishful thinking and his absurd faith in his 'map of islands' constantly excited expectations of landfalls and therefore, indirectly, repeatedly dashed hopes. He welcomed the slightest indication as a sign of nearby land—a chance shower, a passing bird, a supposed river-crab. On 25 September he declared himself certain that his fleet was passing between islands. He did not feel confident enough to turn aside to look for them, though he did venture to inscribe them, with Martín Pinzón's approval, on his chart. On 22 September he was so alarmed by the crews' anxieties that he was glad of an adverse wind. 'I needed such a wind', he wrote, 'because the crew now believed that there were winds in those seas by which we might pass to Spain.'[20]

By the end of the first week in October, when patience must have been at a premium throughout the fleet, Columbus and Pinzón met for an acrimonious interview. So far, with minor adjustments for displacement by adverse winds, they had maintained their westing into the zone where, according to Columbus's calculations, and the speculations of his precious chart, they ought to have found land. Martín Alonso demanded a change of course to the south-west 'for the island of Cipangu', presumably because that was where it was marked on the chart. This suggests that he regarded Columbus's estimate of the distance they had travelled as exaggerated. Columbus refused to comply, on the grounds that 'it was better to go first to the mainland'. He may have persevered in his insistence on an almost unmodified due-westerly course because new discoveries on the latitude of the Canaries belonged, by treaty with Portugal, to the Crown of Castile: on his return, his discoveries were classified, in some reports, as 'new Canary Islands'. His avowed reason, however, suggests that he simply felt a straight course was likely to be quickest and that therefore 'it would be better to go to the mainland first'.[21]

In any event, Columbus's resistance was short-lived. On 7 October, apparently attracted by the direction of the flight of flocks of birds, but perhaps persuaded by an outbreak of mutinous behaviour, he altered course for the south-west. By 10 October, to judge from the paraphrases—which may, of course, have been embellished with hindsight—the men 'could endure no more'. That very night the crisis passed. The following day sightings of flotsam multiplied and as night fell everyone seems to have been excitedly anticipating land. During the night there were a number of claims to have spotted lights from on shore, the earliest of which Columbus ascribed to himself. Las Casas paraphrased the journal: 'the Admiral had it for certain that they were next to land. He said that to the first man to call that he had sighted land he would give a coat of silk without counting the other rewards which the King and Queen had promised.' That night, Columbus thought he discerned a light on the horizon. At two o'clock on the morning of Friday, the 12th, a seaman of Seville, perhaps of Triana, straining from aloft on the *Pinta*, set up the cry of '¡Tierra, tierra!' ('Land, Land!) probably coupled with '¡Albricias!'—the call for rewards. The shot from a small cannon—the agreed signal for land—rang out and was answered on all three ships with praises to God for their answered prayers. Columbus claimed the monarchs' bounty for himself, on the grounds that he had seen the land the previous night, to the unrecorded but presumable chagrin of the look-out from Triana.[22]

An enormous amount of time and effort has been wasted on attempts to identify the island where Columbus made his first landfall. The toponymy of the islands he visited on this first voyage has changed too much, and his surviving descriptions, except of Cuba and Hispaniola, are too vague, inaccurate, and mutually contradictory, linked by sailing directions too corruptly transmitted, to reconstruct his route around them with any confidence. The first island, supposedly called Guanahaní by the natives and named San Salvador by Columbus, was flat, fertile, inhabited, dotted with pools, largely protected by a reef; it was variously described by Columbus as 'small' and 'fairly big', and with what he called a 'lagoon' in the middle; it had a small spit or peninsula at one point on the eastern side and an exploitable natural harbour. Apart from the 'lagoon', the meaning of which is unclear, none of these is a potentially distinguishing feature. As he was approaching from the latitude of Gomera and heading south-west (if what survives of his log can be relied on), Columbus could have arrived at almost any of the islands of the Bahamas or Turks and Caicos, which screen Cuba and Jamaica from such an approach. For what it is worth—and it is not, perhaps, very much—Spanish maps of the early sixteenth century suggest that cartographical tradition came to identify the island of

Columbus's first landfall with that called San Salvador (formerly Watling Island) today.[23]

Two things in particular impressed Columbus when he disembarked to examine the island in the morning light. It struck him as of pleasant aspect, well watered and wooded with an abundance of fruit: he was beholding it, of course, with a paternal and promoter's eye. But before noticing anything about the land—if Las Casas's paraphrase can be trusted—Columbus recorded Europeans' first sight of the natives, whom he called 'naked people'. This was not just a description, but a classification. A late fifteenth-century reader would have understood that Columbus was confronting 'natural men', not the citizens of a civil society possessed of legitimate political institutions of their own. The registering of this perception thus prepared the way for the next step, the ritual appropriation of sovereignty to the Castilian monarchs, with a royal banner streaming and a scribe to record the act of possession. Clothes were the standard by which a people's level of civilization was judged in medieval Latin Christendom. It became an almost frantic preoccupation of Spanish governors early in the history of the New World to persuade the natives to don European dress, just as Spaniards at home had been to much trouble and expense to persuade conquered Moors to 'dress like Christians' and had troubled deeply over the nakedness of the aboriginal Canary Islanders. At a further level, in terms of the two great traditions of thought to which Columbus and his contemporaries were heirs—those of classical antiquity and medieval Christianity—social nakedness might signify either of two conclusions: it could evoke the sort of sylvan simplicity of which classical poets sang and which humanists associated with the 'age of gold'; or it might suggest the state of dependence on God which was starkly symbolized by St Francis of Assisi when he tore off his clothes in the public square. Columbus, whose friends included both humanists and Franciscans, did not explicitly apply either of these paradigms to the natives of the Caribbean but he developed the conclusions which flowed from them: they presented, because of their innocence, a unique opportunity for spreading the gospel; because of their primitivism, an unequalled chance to confer on them the presumable benefits of Latin civilization; and because of their defencelessness, an irresistible object of exploitation.[24]

As he passed through the islands, Columbus was to encounter a variety of indigenous cultures, from the 'backward' world of the Lucayas to the materially rich and technically impressive Tainos of Hispaniola.

But, though he was alert for signs of increasing 'civilization' in his quest for alluring Oriental lands, he saw them all with the same eyes, and all the themes around which he organized his reports are already present in his account of his first encounter on 12 October 1492. First, he constantly compares his hosts, implicitly or explicitly, with Canary Islanders, Blacks, and the monstrous humanoid races which were popularly supposed to inhabit unexplored parts of the earth. The purpose of those comparisons was not so much to convey an idea of what the islanders were really like as to establish doctrinal points: the people were comparable with others who inhabited similar latitudes, in conformity with a doctrine of Aristotle's; they were physically normal, not monstrous, and therefore—according to a commonplace of medieval psychology—fully human and rational. This qualified them as suitable converts to Christianity, which it was a professed objective of Columbus's royal patrons to find.

Secondly, Columbus was anxious to ascribe natural goodness to the inhabitants. He portrayed them as inoffensive, unwarlike creatures, uncorrupted by material greed—indeed, improved by poverty. He ascribed to them an inkling of natural religion, undiverted into what were considered 'unnatural' channels such as idolatry. By implication, they were a moral example to Christians. This picture is strongly reminiscent of a long tradition of late medieval treatment of pagan primitives, especially from Franciscan and humanist writers. The evidence of Columbus's eyes was filtered, in his mind, through expectations derived from tradition.

Thirdly, Columbus was on the lookout for ways of manipulating the natives for profit. At first sight, this seems at variance with his praise for their own moral qualifications; but many of his observations cut two ways. The natives' ignorance of warfare established their innocent credentials but also made them 'easy to conquer'. Their nakedness evoked a sylvan idyll or an ideal of dependence on God, but also suggested savagery and similarity to beasts. Their commercial inexpertise, which made Columbus marvel at the way they exchanged treasures for trifles, showed that they were both morally uncorrupted and easily duped. Their rational faculties made them identifiable as humans and exploitable as slaves. Columbus's attitude was not necessarily duplicitous, only ambiguous; he was genuinely torn between conflicting ways of classifying the Indians. While manipulating learned categories, he frequently slipped into the language of late medieval *mirabilia* or travellers' tales. He singled out whatever struck him as bizarre, funny, quaint, or picturesque. He reported cannibals sceptically, Amazons credulously. Throughout his journeys to the New World he remained divided between rival perceptions of its peoples—as potential Christians, as types of pagan virtue, as exploitable chattels, as figures of fun.

In none of his initial impressions of the New World—neither of the land nor of its people—did he claim to detect any evidence that he was in Asia. The terms in which he saw his discovery seem rather to have been indebted to the experience he had on his West African voyage. He called canoes, for instance, *almadías* and spears *azagayas*—both Portuguese West African terms.[25] It was, however, as if he recovered himself all of a sudden and recalled the task in hand. The search for gold and for Asiatic lands began the day after his arrival, with enquiries after the island of Cipangu.

From 15 to 23 October he reconnoitred three small islands which he called Santa María de la Concepción, Fernandina, and Isabela. He had thus honoured our Lord (at San Salvador), our Lady, and the King and Queen of Spain, in that order. There is not enough reliable information about the islands or about their relationship to one another in the surviving versions of his account for any of them to be securely identified, though the modern Crooked Island—given its size and position—is likely to have been one of them and the modern Long Island was perhaps another. Because of inaccuracies in the original or errors of transcription, many of his sailing directions make little or no sense and the surviving sources are an almost useless guide to his route through the islands.

He felt, however—or, at least, he wished to give the impression—that he was making progress as he sailed among them. The natives, despite their basic similarity one to another, gradually became, in Columbus's eyes, more civilized or more astute. In one place, they knew how to drive a bargain; in another, the women wore a sketchy form of dress; in yet another, the houses were well and cleanly kept. Culled by sign language or randomly interpreted from the utterances of the natives, indications multiplied of the proximity of mature polities, headed by kings. Though we cannot know for certain where to place these islands on the map of the Caribbean, they occupy an important position on the map that was in Columbus's mind: serially aligned, leading towards the imagined 'land which must be profitable'. In the Admiral's imagination, the first big piece of gold reported to him, on 17 October, became an example of the coinage of some great prince; the *Cani*ba, or Caribs, became the people of the Great *Khan*.

For three months Columbus sailed about the Caribbean, dispensing new names to the islands and trinkets to the natives, reading references to the Great Khan or to the land of Cipangu into every garbled native legend or ill-pronounced name that he heard, and always hoping that the next island in the offing might be Cipangu itself. On arriving in Cuba on 24 October he declared, 'I believe this is the island of Cipangu, of which marvellous things are told. And in the globes and painted planispheres I have seen, it is located in this area.' He seems to have realized very soon that this was an illusion, but he discarded it in favour of an even more adventurous assumption: that Cuba could be part of the mainland of Cathay. This suspicion faced him with a dilemma: to continue the search for Cipangu offshore, or to seek the court of the Great Khan in the interior of Cuba. For a time he inclined to the second course, even dispatching an embassy with his 'Chaldean-speaking' interpreter to enquire inland, 'but, finding no sings of organized government, they decided to return'.[26]

Gradually, over the period of his stay in Cuba, Columbus stressed more and more the peculiar merits of a land that could be appreciated for its own sake. The themes of praise for the local environment and enjoyment of its beauty, which had been broached earlier in his narrative, now became dominant. He was preparing for the monarchs— and, perhaps, persuading himself of—a case for the colonization and direct exploitation of the discoveries for their own products, quite apart from any value they might have had as entrepôts for the imagined benefits of Oriental trade. Cuba emerged from his stilted descriptions not as a real place but a literary *locus amoenus*, where nothing was described in detail but everything was of the sweetest and fairest and man is at one with the harmonies of nature. It was a superlative land which excelled the powers of tongue to relate or pen to record. Generally, Columbus confessed his inability to be specific about the produce of the soil. When he thought he recognized mastic, he was wrong. He assumed, however, that all the prolific vegetation must contain much that was marketable.

His treatment of the natives, too, underwent a change, or at least a sharpening of one of his accustomed themes at the expense of the others. As the prospects of exploiting them for gain receded, Columbus's hopes of their evangelization seemed to expand. He formulated a vision of a purified Catholic Church, constructed in the monarchs' dominions, partly with the unsullied raw material with which God had presented them in his discoveries, which would be preserved untainted by contaminating influences. 'And your Highnesses, when their days are accomplished—for all of us are mortal—shall leave their kingdoms in a very peaceful condition and free of heresy and malice, and shall be well received in the presence of the eternal Creator.' This project for an ideal apostolic community in the New World, to which Columbus was often to return, resembled a powerful Franciscan vision that came to spur tremendous missionary efforts in the next century. It grew in Columbus's mind, during the remainder of his life, until it came to form, perhaps, the dominant element in his perception of his discoveries and to inform his sense of his own special dignity and role as the executant of a providential purpose.[27]

Meanwhile, some of his *compagnons de voyage* were losing patience with the poor pickings of his discoveries and the rich exhalations of his spirit. On 20 November Martín Pinzón sailed off without leave on an island-hopping gold hunt—either sated with the beauties of Cuba or frustrated by its poverty. After this breaking of ranks, denounced by Columbus as 'treachery', the other explorers were unlikely to endure Cuba for much longer and on 23 November Columbus began to look for a wind to bear him away. When at last he found it, on 5 December, it blew him by chance, by a sudden change of direction, to the most important island he was ever to find. The native name for it was Haiti, but Columbus, having already used up the names of the members of the Spanish royal family, and of the more important of the Holy Family, honoured the nation that patronized his enterprise by calling it La Isla Española, or Hispaniola, as English maps render it. Hispaniola was a significant find for two reasons. In the first place, though it proved not to be Cipangu, it produced fair quantities of gold: this was the making of Columbus's mission; without it, he would almost certainly have returned home to ridicule and obscurity. Secondly, it housed an indigenous culture of sufficient wealth and prowess to impress the Spaniards; with some of the natives Columbus was able to establish friendly relations—as he thought—and to fix in their territory, the intended site of a future colony. In what survives of his account, Columbus made no mention of the superior material civilization of the island: the elaborate stonework and woodwork in ceremonial spaces reminiscent of Mesoamerican ball-courts; the stone collars, pendants, and stylized statues; the richly carved wooden stools known as *duhos*; the anthropomorphic or zoomorphic personal jewellery of gold called *çemis*. Nevertheless, he displayed a distinct awareness that Hispaniola was his best find so far, with the most promising environment and the most ingenious inhabitants.

After extensively reconnoitring the northern coast, Columbus made his first contact with a local polity of Hispaniola near the modern Port Paix. He showed the *cacique* or chief whom he met there due honour and

told him how he came from the monarchs of Castile, who were the greatest princes in the world. But . . . the other would only believe that the Spaniards came from heaven and that the realms of Castile were in heaven . . . All the islands are so utterly at your Highnesses' command that it only remains to establish a Spanish presence and order them to perform your will. For I could traverse all these islands in arms without meeting opposition . . . so that they are yours to command and make them work, sow seed and do whatever else is necessary and build a town and teach them to wear clothes and adopt our customs.[28]

In the changed perceptions of his discoveries revealed by Columbus all the agonies of Spain's dilemma in the New World were foreshadowed. The quick pickings he began by seeking—the exotic products, the commercial gain—had been crowded out of his mind by the spectacle of a sundowner world, imagined complete, with complaisant natives and all the comforts of home. But this unequal Arcady could be constructed only by the exercise of social and moral responsibility: the Indians would have to be 'civilized' in a metropolitan image and the colonists would be teachers as well as masters. There were three distinct projects for future conquests here: the rapid, irresponsible turnover; the long graft of a colonial utopia; the 'civilizing' mission. The Spaniards could suck like leeches, build like bees, or spread an inclusive web like spiders. Neither Columbus nor any of his successors ever resolved the inherent contradictions.

He began to lay the foundations for a colonial empire when he established a strong personal link, enriched with gifts and fortified by the impression made by Spanish weapons of fire and steel, with the most important *cacique* he could find, named Guacanagarí—chief, so it seemed to Columbus, of the northern part of the island. By arrangement with Guacanagarí he took the first step towards establishing the Spanish 'presence' he had set his heart on by erecting a stockade at Puerto Navidad on the north coast and garrisoning it with thirty-nine men who were to remain collecting gold-samples and await a new expedition from Castile.

The establishment of a garrison was genuinely a new departure: it was not inconsistent with Columbus's mission, but it was a sign of a policy not previously envisaged. It can be understood satisfactorily as a development arising from the new opportunities, and from the disappointed expectations, which Columbus encountered and registered. But at the moment when the decision was made, Columbus chose, in a way that henceforth was to be characteristic, to represent it as the result of a sudden revelation, implicitly providential in origin. It happened, according to him, like this. At midnight on Christmas Eve, the *Santa María* ran irrecoverably aground. At first (if Columbus's two mutually contradictory explanations of the disaster can be so combined) Columbus was inclined to blame a lazy seaman who, against orders, left a boy in charge of the tiller. On reflection, the next day, he saw the outcome rather differently, as the result of the treachery of the 'men of Palos' who had begun by providing a dud ship and ended by failing to ease it off the rocks. He was taking refuge, yet again, in a self-identification which is now familiar to the reader: the loner in adversity, the victim of conspiracy. The universal paradigm of treachery—the Judas kiss—demonstrates, however, for any Christian so inclined, that even betrayal has a place, a key place, in the cosmic scheme. For Columbus, the wickedness of his crew was providentially ordained, as surely as that of Judas. 'It was a great blessing', he wrote, 'and the express purpose of God that his ship should run aground there in order that he should leave some of his men'; that it was provoked by treachery was, Las Casas reported, precisely Columbus's proof that the hand of God was behind it. From that open hand, Columbus now received as if by a miracle 'planks to build the fort with and stocks of bread and wine for more than a year and seed to sow and the ship's boat and a caulker and a gunner and a carpenter and a cooper'. The debris of the ship and the dregs of the crew would supply the needs of the moment. Columbus had evidently decided that he wanted to establish a garrison. By this device, he shifted responsibility for the decision to God.[29]

The loss of the *Santa María* turned Columbus's thoughts homeward. He claimed to want to explore further, but his delay was more probably caused by a preoccupation prominent in Las Casas's paraphrases: to accumulate a large quantity of gold. Columbus had gathered small amounts by way of truck from 12 December onwards but was at first disinclined to believe they originated on the island. Numerous presents of gold artefacts in the last fortnight of December, however, seem to have convinced him that the source of the gold was close at hand and in the early days of January the evidence for an elusive 'mine' grew encouragingly. On 6 January Martín Pinzón rejoined the expedition, bearing a further large quantity of gold, which he claimed to have obtained by trading. He presented excuses for his conduct, which

Columbus privately dismissed as lies designed to cloak avarice and pride and even peculation; but the Admiral refrained from outright denunciation 'so as to give no leeway to the evil works of Satan, who desired as much as ever to impede that voyage'.[30] No student of Columbus will be convinced by his self-righteous assertions, nor see in the trajectory of his relationship with Martín Alonso anything other than a typical tale of Columbus's friendships. He was incapable of moderation in personal relationships and we repeatedly see his malice and scorn flare out at the former friends of his bosom. This is not to say that it was Columbus's fault alone that he made rivals and enemies of a series of close collaborators: Martín Pinzón, Amerigo Vespucci and, as we shall see, Juan de la Cosa, Francisco Roldán, Pero Niño, Alonso de Hojeda; rather that his love and affection, often freely given, were always easily dispelled; his trust was confided in its entirety and could not be diminished without being destroyed.

Though indignant with Pinzón, Columbus seems genuinely to have been satisfied with the achievements of the voyage. Though apprehensive at having found no trace of China or Cipangu, he was convinced that in Hispaniola he had an exploitable asset. He claimed to believe the island was larger than Spain and wrote glowingly to the monarchs of the natural virtues of 'the best land in the world'. Events were to prove him false. The climate he praised was to prove lethal to Spaniards. The natives he characterized as peace-loving and biddable turned out to be shifty and deadly. But, on the basis of what he had seen up to the time of his departure, Columbus's assessment, if allowance is made for the exaggerations of a salesman on commission, was justifiable.

He had collected many samples of gold, pods of pungent chili, a little inferior cinnamon—which perhaps aroused hopes of richer spiceries—rumours of pearls, and some human specimens in the form of kidnapped Indians to show off back at court. He had discovered the pineapple, tobacco—'some leaves which must be highly esteemed among the Indians', though he did not yet know what it was for—the canoe, and the hammock, a gift of Caribbean technology to the rest of the world which in years to come was to do much to ameliorate the sleep of his fellow-seamen. He departed on 16 January 1493, a little more than a year after he first received his commission from the monarchs, in an optimistic mood, reflecting that, if Hispaniola was not Cipangu, it was at least 'a marvel', perhaps the realm of Sheba, or the country from which the Magi bore their gifts to Christ.[31]

*　　*　　*

At first, all went well. Dismissing the crew's misgivings, Columbus struck north for the westerlies and found them on 5 February; even before that, he had pretty fair weather and was able to make some easting almost every day. The remainder of the voyage, however, was to be as chequered with disaster and triumph as any comparable period in Columbus's life. It is important to try to picture his state of mind.

He had been through an extraordinary experience, which was almost bound to change him or, at least, to accentuate some existing traits. There are unmistakable signs, in what he wrote on his way home, that Columbus's notion of reality and grasp of the limits of the possible had been deeply shaken by his contact with the New World. When, for example, he had first heard of the cannibals, he dismissed them as a myth, perhaps because he recognized them as a commonplace of medieval imaginative travel-literature. By the time he had seen the wonders of the New World he had changed his mind and was disposed to accept such stories. Tales, perhaps originated by Martín Pinzón, of islands respectively of Amazon women and bald men, no longer aroused his scepticism. His tendency to ascribe changes of fortune to the intervention of supernatural agents had come almost entirely to displace rational forms of explanation. The way the environment of his discoveries defied his understanding—presenting him, for example, with all that bafflingly unclassifiable vegetation, traducing the expectations aroused by his sea-chart, surrounding him with the babble of unintelligible guides—helps to explain the effect.

So does the isolation he endured. In part, what emerges from his account is the sense of collective isolation shared by the whole expedition, whose members doubted whether they would ever get back to Spain. Columbus's further sense of personal isolation was exacerbated by his own uncompanionable temperament and his peculiar circumstances. He suffered the loneliness of command. He was 'a foreigner' whose habits and interests not only isolated but alienated him from his men. He belonged to none of the almost ethnic cliques of which the crews were composed: the Basques, for instance, who rioted together, or the men of Palos and Moguer who were the friends and employees of Pinzón. He faced the unremitting fear of mutiny or perfidy—and fear is always real to its victims, even if the grounds for it are false.

It is not surprising that in this state of enforced and fragile self-reliance, in the mood of exaltation induced by the revelation of so many of the 'secrets of this world', Columbus should have turned to God. Religion was always his first refuge in adversity. The piety of *The First*

Voyage may be misleading: the editorial hand of Las Casas highlights every reference to God, but there is an unmistakable regularity about the readiness with which Columbus reaches for the consolations of faith. When disaster first loomed, in mid-September 1492, he responded by comparing himself with Noah and Moses. In early November, when he began to despair of finding anything commercially useful, he extolled the prospects of profit for the soul. When catastrophe stranded the *Santa María*, Columbus explained it away as a miracle. When he fell out with Pinzón, he blamed the Devil. On his way back to Spain, he was ready for the most acute and profound religious experience he had yet recorded: the first of a series of mystical experiences that were to mark a long and sometimes precipitate spiritual progress towards the intense religiosity of his later life.

When it happened, on 14 February 1493, he was literally lost. He thought he was well south of his true position—nearer the Canaries than the Azores; his uncertainty was shared by the professional pilots aboard, who pronounced themselves effectively helpless. Danger was added to insecurity when they ran into a terrible storm, which sundered the ships and put every man in fear of his life. Columbus's own thoughts in the throes of the storm are captured in a fragment of his own writing, transmitted by a late and corrupt source but resonant with credibility:

I could have endured this raging sea with less anguish if my person alone had been in danger, for I knew that my life was at the disposal of Him Who made me; and I have been near death so often, and so close that it seemed that the best step I could take was the one that separated me from it. What made it so unbearably painful this time was the thought that after our Lord had been pleased to enflame me with faith and trust in this enterprise, and had crowned it with victory, so that my foes would be humbled and your Highnesses served by me to the honour and increase of your high estate, His divine Majesty should now choose to jeopardize everything with my death. I could have succumbed happily but for the danger to the lives of the men I had taken with me on the promise of a prosperous outcome. In their terrible affliction they cursed their coming and regretted that they had let me cajole or coerce them into sailing on, when they had so often wished to turn back.

Then was my anguish redoubled, for I seemed to see before my eyes the remembered vision of my two boys at school in Cordova, abandoned without help in a foreign land, before I had accomplished for your Highnesses the service that might dispose you to remember them with favour—or, at any rate, before I had made you aware of it. And I tried to console myself with the thought that our Lord would not allow such an enterprise to remain unfinished, which was so much for the exaltation of His Church and which I had brought to pass with so much travail in the face of such hostility, nor would He want to break me; yet I realized that He might choose to humble me for my sins, to deprive me of the glory of this world.[32]

This impressive language enshrined a good deal of bunkum: Columbus had not succeeded in his mission—which had been to find a route to Asia—though he might be pardoned the illusion; his concern for his crew, which can be assumed in any responsible sea-captain, had never been mentioned in any lesser adversity and contrasts with his previous denunciations of their treachery; nor had Columbus's voyage been undertaken for the Church: that motif had first arisen as a ploy of royal propaganda and was later developed by Columbus to make up for poor returns in other respects. Yet Columbus's words have the ring of a death-bed confession: it would be rash to dismiss them as insincere.

At this point, when he already had one 'vision before my eyes', Columbus seems to have experienced another apparition, in the form of a voice, of apparently celestial origin, pouring religious solace into his ear. This was the first, but not the last, time this voice came to comfort him in affliction. Its words, which survive in Las Casas's version only in reported speech, are not on this occasion expressly attributed to the voice, but their content is recognizable from its later reappearances. It summarized the mercies God had conferred on him; made an implicit and unflattering comparison between Divine generosity and royal parsimony; confirmed Columbus's contempt for those who had opposed his project; recalled his 'trials and adversities' and assured him that all were trials of faith, of little account by comparison with 'the things of great wonder which God had performed in him and through him in the course of that voyage'.[33]

In a letter apparently written aboard ship the following day—but perhaps touched up by an editor with a view to publication—Columbus summed up his achievement with due praise to

our Lord God everlasting, Who gives to those who walk in His way the conquest of what appears impossible. And this, conspicuously, was just such a conquest, for although these lands may have been inferred and written about, it has all been speculative up to now, without confirmation by sight, without full understanding—so much so that most of those who heard about them listened and adjudged them more likely to be legendary than anything else.

With suppression of Columbus's usual egoism, the letter goes on to attribute the discovery thus ringingly proclaimed to the monarchs of

Castile, presumably in an attempt to stake a pre-emptive claim to sovereignty:

And so it is that our Redeemer has granted to our most illustrious King and Queen, and to their famous kingdoms the achievement of so lofty a matter, at which all Christendom must rejoice and celebrate great festivities and give solemn thanks to the Holy Trinity, with many solemn prayers, for the exaltation that shall be derived from the conversion of so many peoples to our holy faith and, secondly, for the material benefits which will bring refreshment and profit.[34]

Not even celestial voices could ever quite distract Columbus from the main chance.

The *Niña* emerged safely from the storm into the harbour of Santa Maria in the Azores on 18 February 1493. The *Pinta* had disappeared. The Atlantic crossing had barely been completed. The Portuguese authorities on the island had no welcome for Columbus and his men, not because they were jealous of the extension of Castilian power which Columbus had just effected, for they probably knew nothing of it and would not have credited it had they been told, but rather because relations between Spain and Portugal were generally bad and Castilian vessels in Portuguese waters were automatically under suspicion of piracy. Ten men who went ashore to offer prayers to the local Virgin on their deliverance from the storm were clapped in irons, and Columbus had much difficulty in extricating them. He still had a long run back home. He had lost the *Pinta*, it seemed, at sea, as well as the *Santa María* in Hispaniola. The weather was still squally and threatening.

An ill wind now bore him from the Azores through more storms to Lisbon. His old correspondent, King João II, was not as simple as the officials in the Azores. He knew Columbus's businesses in the Ocean Sea and was displeased that a voyage he had refused to sponsor had turned out promisingly under the auspices of his rivals. It may therefore have been with some trepidation—but no option, in view of the state of his vessel and crew after such a taxing journey—that Columbus disembarked in the Portuguese capital. Not only was he apprehended by King João and left in doubt of his chances of release, but he also came under suspicion in Castile for this unexpected—but probably involuntary—intercourse with an enemy. Columbus was not above threatening to transfer his services; and the example of his fellow-Genoese, Antonio da Noli, who had shifted allegiance between Castile and Portugal during the war of 1474–9, gave his countrymen a poor reputation for loyalty. It is hard to imagine, however, what Columbus might have gained from defection at this juncture, unless all his confidence in his discoveries were sham—surely an unlikely hypothesis, even for someone with Columbus's dissembling talents.

King João, indeed, seems already to have been contemplating a diplomatic arrangement that would include concessions to Castile in the Ocean Sea in return for an indisputably Portuguese zone around southern Africa. Columbus was released and in mid-April 1493, 'Don Cristóbal Colón, Admiral of the Ocean Sea, Viceroy and Governor of the Islands he has discovered in the Indies'—as he was now deservedly, if misleadingly styled—could dangle his gold-samples and parade his feathered Indians before the admiring court of Ferdinand and Isabella in Barcelona.

Before he did so, one more extraordinary stroke of fortune was to befall his enterprise. The *Pinta*, with Pinzón on board, had escaped from the February storm, though separated from the *Niña*, and had struggled into the northern Spanish port of Bayona ahead of Columbus. Martín Pinzón had every motive, and some reason, to mount a challenge to Columbus's claims. Later, during the long lawsuit between the Crown and Columbus's family, friends of the Pinzóns constructed an elaborate legend around Martín's memory, attributing to him the major role in the great voyage of discovery. Had Martín given the monarchs his own account of events, Columbus might have been severely embarrassed, his glory divided, perhaps, his career curtailed. As it happened, there was no one, save the troupe of captive Indians, to share the stage with Columbus in Barcelona. For, taxed beyond his strength by the sleepless weeks in a stormy sea, shortly after finding a haven in Castile and before he had time to tell his own story, Martín Pinzón was dead.

'Your More Divine than Human Journey'

MARCH 1493–JUNE 1496 AND THE SECOND CROSSING

'LIFT up your hearts!' wrote Peter Martyr of Anghiera, relaying Columbus's tidings to his correspondents, 'Oh, happy deed, that under the patronage of my King and Queen the disclosure has begun of what was hidden from the first creation of the world!' The means that took Columbus across the Ocean Sea struck one respected cosmographer as 'more divine than human'.[1] How Columbus and the news he brought were received at home is germane to a question we shall soon have to face: whether or in what sense Columbus can be said to have 'discovered' America. The initial euphoria masked searching questions about the nature of his finds and the status of his promises.

His own professed belief was that the lands he had discovered were Asiatic; yet he was equivocal enough also to call them 'unknown'. He was capable of admitting other possible identifications to himself and to the monarchs: on one later occasion, as we shall see, he actually did so; but in the face of contradiction by others he never wavered in his insistence that his promise to find a western passage to Asia had been fulfilled. Almost from the moment of his return from his first voyage his burden became, 'I have found and continue to find nothing less in any respect than what I wrote and said and affirmed to their Highnesses in days gone by.'[2]

This characteristic wording discloses the motive for his obduracy: one of the focuses of his opponents' attacks was that his discoveries had not represented performance of his bargain with his patrons. He had not found a way to Asia, but to a group of islands similar to those already known or, perhaps, proximate to an antipodal land. Columbus's safe and surprising return; the letters he sent to correspondents at court and in Cordova; his spectacular presentation in Barcelona, with an exotic parade and tastings of samples of piquant condiments; and the rapid circulation of his printed report, which had its first edition in Barcelona even before Columbus got there: these eye-catching events—as intellectually intriguing as they were sensually stimulating—immediately aroused advocates of all three possible theories of the nature of the newly found lands: those who agreed with Columbus; those who classified his discoveries as Antipodean; and those who regarded them simply as more islands of the sort already known.

From Columbus's point of view, it was important that Ferdinand's and Isabella's opinion should coincide with his own. The monarchs' first reaction was to accept the authenticity of his claims, but they were not sufficiently confident of him to commit themselves irrevocably. In his summons to court, Columbus was addressed with all the titles he had been promised upon the successful completion of his enterprise, but the lands of his recent adventures were identified only vaguely as 'islands he has discovered in the Indies'. By August 1494, the monarchs were more thoroughly persuaded: 'it seems,' they wrote to Columbus, 'that everything which from the first, you said could be achieved, has turned out for the most part [sic] to be true, as if you had seen it before you spoke of it.' Royal enquiries at that time about whether the seasons in the new lands coincided with those at home are reminiscent of some remarks about the Antipodes attributed to Posidonius—but the matter is obscure, and other writers had connected such seasonal variations with oriental climes.[3]

In the negotiations which were soon opened with Portugal for the confirmation of Castilian sovereignty in these areas, the monarchs' servants used even more imprecise language. If anything, the royal and papal chanceries at the time of the drafting of the Bull *Inter Cetera*, published shortly after Columbus's return, inclined to the view that the explorer had found an antipodal continent, for the phrase 'mainlands and islands remote and unknown' which they applied to the discoveries appeared to exclude Asia, which was not 'unknown' in a sense current in circles exposed to classical learning, but merely long unvisited. The name of 'Antipodes' was actually bestowed on Columbus's islands in one of the first reports to leave court after his arrival there, written by Peter Martyr of Anghiera, for the information of friends in Italy: 'there has returned from the western Antipodes one Christopher Columbus of Liguria, who barely obtained three ships from my sovereigns for the voyage, for they regarded the things he said as fabulous.' To other correspondents, Peter Martyr declared that Columbus's discoveries were previously unknown—by which again he meant that they were not Asiatic. Whereas he normally used the names of 'Antipodes' and 'New World' or 'New Orb', he was always non-committal or positively hostile whenever relaying Columbus's own opinion that he had sailed to regions near India. Peter Martyr's opinion seems to have been the dominant one among the Italian humanists. In a sermon at Rome in 1497 one of them described how Columbus had taken the Name of Christ to the Antipodes 'which previously we did not even think existed', and not long afterwards another in Florence designated Columbus's discovery as 'the other world opposite to our own'.[4]

Other conflicting views were current. Immediately on Columbus's arrival in Lisbon, it was rumoured that he had found the mythical lost land of Antillia—which, if true, would have been to Portugal's advantage for legend ascribed Antillia to Portuguese founding-fathers. Some early reports in Italy and Castile characterized the finds as new Canary Islands—which was not unreasonable, since they were on about the same latitude and seemed to evince some cultural similarities. As the Canaries were guaranteed by treaty to Castile, this was as politically interested an identification as that of Antillia. Both rumours associated Columbus's explorations with areas that were neither Asiatic nor Antipodean in character. Most reports designated the discoveries as 'islands' without any more detailed commitment.[5]

Despite such rumours and the views of Peter Martyr and his friends, while many scholars suspended judgement, other men were drawn to share Columbus's belief that the new lands were part of Asia. The explorer's own apparent assurance—which had gained a certain stature with his safe return—and the undoubted samples of gold he carried helped to create this impression. The Duke, formerly Count, of Medinaceli asked the monarchs of Spain for permission to exploit the good fortune of his erstwhile protégé, who had 'found all he was looking for', by sending caravels to trade for spices. Various Italians, sometimes drawing for their information directly on the printed version of Columbus's account, reported home to the same effect; their opinion does not seem to have carried as much weight as Peter Martyr's, though the Duke of Ferrara accepted it and supposed that Toscanelli's theories must have had some relevance to Columbus's success. Scholars generally continued to adhere to traditional estimates of the size of the globe and therefore could not agree that Columbus could have reached Asia. Peter Martyr thought the size of the world precluded it. Jaume Ferrer, the Majorcan cosmographer, realized that Columbus's world was underestimated. As the explorer's friend, Andrés Bernáldez, told him frankly, he could have 'sailed another twelve hundred leagues' and still not have got there, though he did accept that 'the land was continuous' from Columbus's discoveries westwards across the island-earth. Had a western route to Asia really have been thought possible in Italy where economic life depended so heavily on established patterns of long-range trade, there would have been an upheaval in the markets and a seismic upsurge of diplomatic activity. It is true on the other hand that in the next pontificate, that of Julius II, the papal chancery (probably, as usual, merely adopting the language of the petitioner) conveyed the impression that the Columbine discoveries were Oriental, since a Bull of 1504 located unspecified conquests by the monarchs of Spain 'in parts of Asia' and went on to establish three new sees on the island of Hispaniola. Explorers of the New World in the early years of the sixteenth century, including Vespucci and Vicente Yañez Pinzón, made their voyages on the assumption that Columbus was roughly right and that his route led to, or near to, Asia. The issue was unclear. Yet it is evident that on Columbus's first return, the minds of men in the Old World were rapidly able to adjust to the idea that a continent such as America existed—a new world, different from the known landmass—and had already anticipated its discovery under the name of 'the Antipodes'.⁶

Without committing themselves to the view that Columbus had visited Asia, Ferdinand and Isabella were sufficiently taken with his gold-samples and his specimens and accounts of the natives to regard the discovery as of serious importance. The honour they showed him, as well as their eagerness to invest in further exploration of the area, is proof of this, for they allowed him to sit in their presence and ride beside them at ceremonies or in procession. Columbus entered a new phase of his career: years of acclaim, in which his detractors were obliged to bide their time while his admirers lionized him. For Peter Martyr, he was like one of the heroes of whom the ancients made gods. For Jaume Ferrer, he resembled an apostle, performing for the West what St Thomas had done for the East.⁷ Thus the fugitive from the weaver's shop in Genoa and the tavern in Savona was elevated to the highest ranks of pagan and Christian hero-worship. It was, no doubt, a gratifying role, but one Columbus was ill equipped to sustain for long, especially as it brought him new responsibilities, way beyond his competence, as a cosmo-graphical adviser to the Crown, a diplomatic consultant for negotiations with Portugal, and a colonial administrator in the empire he had begun to found.

The international negotiations were the most delicate part of the preparations involved in the next stage of the discovery and exploitation of the Indies. The Spaniards' aim was to agree a line of demarcation in the Ocean Sea, beyond which all new lands would be assigned to Castile. The Portuguese, travelling eastwards from such a line, would take all the lands they found in their explorations eastward round Africa, until they met the Castilians, who would be navigating from the west. The effect of this, in the Spaniards' estimation, would be to ensure the whole Orient for themselves: it was not for another four years that the Portuguese mastered the difficulty of the route to Asia round the Cape of Good Hope. The so-called *Memoria de la Mejorada*—a memorandum on the subject addressed to Ferdinand and Isabella and recently, but riskily, attributed to Columbus—suggests that the Cape of Good Hope be treated as the complementary line of demarcation dividing Castilian from Portuguese lands in the east, and lists India, Persia, Arabia, and East Africa as conquests to be adjudged to Spain.[8] During 1493 only the question of the western boundary seems to have been raised.

While Columbus was involved in the negotiations, his own desire prevailed for a line running from north to south a hundred leagues west of the Azores, where he thought he had detected a change of clime during his voyage, to a more pleasant and mellower atmosphere. The Spaniards obtained papal agreement and a confirmatory Bull before Columbus's departure on his second voyage, but the Portuguese were unwilling to confine their westward navigations within such strict limits: in the following years their ships would push far into the Atlantic in order to get the benefit of the north-east trades in an attempt to make enough southing to round the Cape of Good Hope. The state of play by the time of Columbus's departure in September 1493 was expressed in a letter of Isabella's to him, which also reveals the relations still obtaining between Columbus and his sovereigns at the time:

Don Cristóbal Colón, my Admiral of the Ocean Sea, Viceroy and Governor of the islands newly discovered in the Indies. With this messenger I send you a copy of the book which you left here, which has been so long delayed because it has been made secretly so that the Portuguese emissaries here should not know of it, nor anyone else; and for the same reason it has been done in two hands, as you will see, for the sake of speed. Certainly, according to what has been said and seen in the present negotiations here, we know increasingly from day to day the importance, greatness, and substantial nature of the business, and that you have served us well therein; and we place great reliance on you and hope in God that beyond what we have promised you, which shall be most fully met and honoured, you will receive from us much more honour, grace, and increase, as is right and as your services and merits deserve. The sea-chart which you have to make you will send me when it is finished; and to serve me you will make great speed in your departure so that, if our Lord is gracious, the chart may be commenced without delay, for you must see how important it is to the progress of the negotiations. And of all that happens at your destination you will write and always let us know. In the Portuguese negotiations nothing has been decided with the envoys who are here, although I believe their King will come to see reason in the matter. I could wish you thought otherwise, so that you would therefore not delay but proceed at once to the task in hand, to avoid any possibility of false hopes.[9]

The Queen's anxiety for a 'chart' reflects, it is safe to suppose, the difficulty of establishing where, in practice, any line of demarcation fell. Columbus's promise to make a grid map of his route and discoveries, with the co-ordinates shown, would, if fulfilled, resolve the problem; but he had not done it, and never would, because it was beyond the technical capacity of the time. The theory of longitude—that it could be fixed by calculating the difference in time of a predictable celestial occurrence between a chosen place and a standard meridian—was well known. But there were no known suitable phenomena, other than eclipses, until in the next century the moons of Jupiter were revealed by telescope; and there were no means of keeping time with sufficient accuracy. Columbus's later efforts to use eclipses were wildly inaccurate. During the negotiations with Portugal, the monarchs consulted Jaume Ferrer about the problem. He lamely suggested that the distance would just have to be measured after the traditional methods of mariners—which were admittedly hit-and-miss—and referred his correspondents to Columbus.[10]

The progress of negotiations over the following months was shown by Isabella's next communication to Columbus on the subject, written in August 1494 when the recipient was back in Hispaniola:

Since matters with Portugal are now agreed, ships can come and go in perfect safety. . . . An arrangement was made with my ambassadors and on the question of the demarcation line or boundary which has still to be made, because it seems to us a problem of great difficulty, we should like you, if possible, to play a part in the negotiations. . . . See whether your brother [Bartolomé] or anyone else you have with you can master the question. Brief them very fully orally and

I. First printed in Barcelona in April 1493, the supposed first report of Columbus's discovery sold so well that illustrated editions printed in Basle rapidly followed. The artist's impressions, though highly fanciful, were based on close reading of the text. Here, the first four islands mentioned are stylized, with the fair hills, woodlands, and 'great cities and towns' Columbus attributed to them. Even the lone figure manipulating the rigging (if intended for Columbus) illustrates an element of the explorer's self-image. The impressive cityscape at bottom left may have been inspired by the *Carta*'s claim that Columbus had captured a large town 'and there have I built a great stronghold and fortress'. The prominence given to islands ('Fernada' and 'Ysabella') named after the Castilian king and queen illustrates a strong theme of royal propaganda in the *Carta*—the pre-emptive assertion of a Castilian claim to sovereignty over the discoveries. See pp. 92–3.

II. Although no galley could have made a crossing of the Atlantic, the artist of this woodcut from a Basle edition of the *Carta* chose his image to suggest the lucrative commerce which, the *Carta* claimed, could be found in Columbus's discoveries: 'in this Hispaniola, in the most suitable place and the best area for the gold mines and for trade, both with the mainland on this side of the ocean, and that on the far side belonging to the Great Khan, great trade and profit will be had.' The exploitability of the natives was emphasized: 'Whatever thing they have, if you ask for it, they never say no.' The artist also captures the *Carta*'s insistence on their 'incredible' and 'irremediable' timidity. The figures in oriental headgear in the woodcut are presumably meant to represent merchants from a supposedly nearby Asiatic civilization, perhaps subjects of the Great Khan himself. The trees recall the *Carta*'s assurances of 'marvellous pine trees'. There is an odd romantic touch, reminiscent of Flemish landscape painting of the time, in the fantastic rocks in the background.

III. This illustration to the *Carta* shows, in the prow of the ship, the earliest known attempt unequivocally to portray Columbus. The island-scene exactly illustrates the text: 'The people of [Hispaniola] . . . all go naked, . . . although some of the women cover themselves up in just one place with a leaf of grass or a slip of cotton . . . They carry no arms except for staves . . . Many times I happen to have sent two or three men to some village and a countless horde of the inhabitants has come out and all have run away.' Palm trees 'which it is a wonder to see' are also in the text. In the background, unrelated to the text, evocations of native *bohíos* are close to early drawings from life. They contrast with the sophisticated buildings in earlier woodcuts. The majestic King Ferdinand illustrates the *Carta*'s role in advancing Castilian political claims. Cf. plate I and p. 93.

IV. Made in Nuremburg in 1492, Martin Behaim's is the oldest surviving globe of the world. Some months after Columbus had returned from the New World, Behaim arrived in Lisbon, bearing a letter from his fellow-citizen Hieronymus Müntzer, which proposed a transatlantic expedition in search of a short route to Asia. 'And what glory will you gather if you make the habitable east known to the west and make those eastern islands subject to you!' The globe shows Behaim to have been a small-world theorist. Like Columbus, he reduced the Ocean Sea to navigable proportions, used Marco Polo and John Mandeville, and emphasized Cipangu, praised as 'the most noble and richest island of the east, full of spices and precious stones' as well as gold. Columbus's mental map seems to have differed slightly, making the world of even smaller proportions and locating Cipangu further south. See pp. 29–30, 80 and map 2.

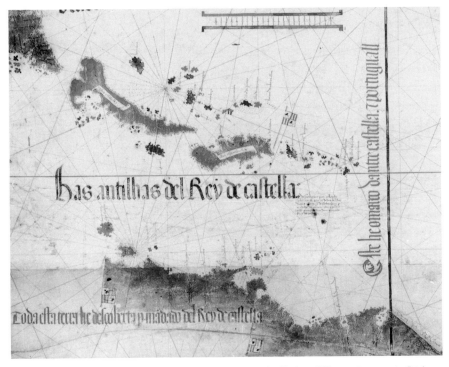

has antilhas del Rey de castella:

@ he he ornano santa castella. 7 portugual

Toda esta terra he descoberta p mãdado del Rey de castella

V. In this world map purchased by Alberto Cantino, the Duke of Ferrara's agent in Lisbon, before November 1502, the chief purpose was to portray 'the islands recently discovered in the regions of the Indies'. The Tordesillas line—drawn by treaty in 1494 between zones of Castilian and Portuguese expansion—forms the thick rib from north to south, with Brazil in roughly the right relation to it. The Caribbean is marked with the legend, 'The Antilles of the King of Castile, discovered by Columbus the Admiral, as he is, of those said islands, which were discovered by order of the very high and mighty prince, King Ferdinand, king of Castile'. The depiction of the Lesser Antilles, Hispaniola and Jamaica reflects Columbus's reports. Cuba is shown twice, as both island and continent (unless an unrecorded discovery of Florida or Yucatán is postulated to account for the peninsula on the left). More Bahamas are shown than Columbus knew at first hand. See pp. 109–10.

VI. Like the Piri Re'is map (plate VIII), that signed by Juan de la Cosa shows the
New World as a continuous land-mass but smothers the Isthmus, where Columbus sought a strait on his last voyage, with a large image—too worn for reproduc-
tion here—of St Christopher bearing the Christ-child on his back, perhaps in
allusion to the Columbus's self-appointed role as 'Christo Ferens', or 'Bearer of
Christ'. Though various wind-roses are scattered over the map, that of the Virgin
is the largest and most prominent, set athwart the Tropic of Cancer in a central
position. The style resembles that of a woodcut, from which she can be presumed
to have been copied. Though shown with the infant-Christ, and attended by
angels, rather than crowned and accompanied by the persons of the Trinity, her
presentation evokes the sort of altar-pieces presumed to have been represented
by Columbus's mystic signature. See pp. 118–19.

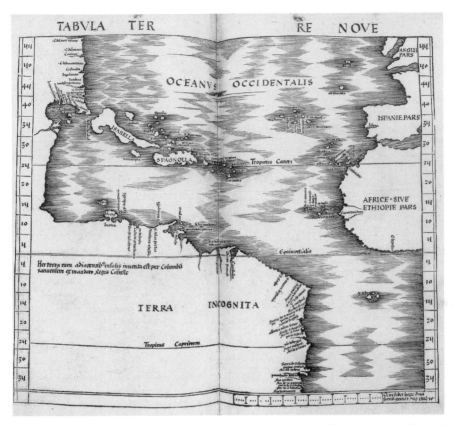

VII. When Martin Waldseemüller published his vast world-map to illustrate a new edition of Ptolemy in Strasburg in 1507, he gave pride of place to Vespucci, depicting him in a presiding position, as discoverer of the New World, in place and posture equipollent with Ptolemy himself. The *Tabula Terre Nove*, published in 1513, redressed the injustice by inserting the legend (visible on the left): 'This land with the adjacent islands was discovered by Columbus the Genoese by order of the King of Castile.'

Unlike the maps of Cantino, de la Cosa and Piri Re'is (plates V, VI, and VIII), this has no flavour of a marine chart. Instead of the delicate web of rhumb-lines, designed to provide sailors with direction-finding bearings from point to point, this map substitutes a 'scientific' method of arrangement, working towards the grid of co-ordinates proposed by Ptolemy. Cf. p. 76.

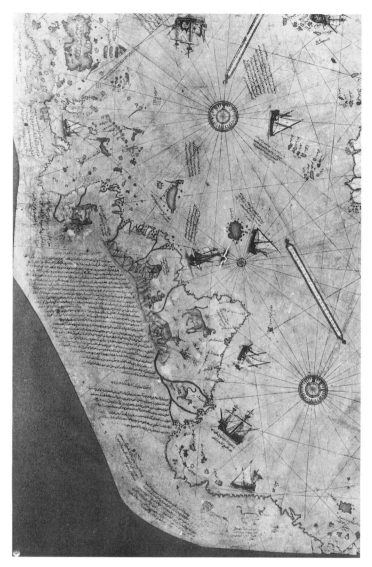

VIII. Ottoman reaction to Columbus's discoveries is exemplified by the work of Piri Re'is of 1513. He claimed that among the sources of his map were charts made by Columbus himself and captured at sea by Kamal Re'is in 1501. The long note on the left, near the neck of the parchment, summarizes Columbus's first ocean crossing. The depiction of the lesser Antilles recalls Columbus's second voyage and resembles Juan de la Cosa's map of a few years previously. Piri Re'is astonished his fellow-countrymen by copying the conventions, both cartographic and pictorial, of his Christian contemporaries. The detail of campers on a whale's back illustrates an episode of the legend of St Brendan. See p. 36.

in writing and perhaps with a map . . . and send them back to us in the next fleet.[11]

The agreement Isabella referred to was the Treaty of Tordesillas of June 1494, in which the Portuguese agreed to a demarcation line in principle and fixed the western limit at a point 370 leagues beyond the Cape Verde Islands—a decision rather more favourable to themselves than that made by the Pope and one which eventually secured much of Brazil for Portugal. The limit which still had to be made was the eastern one. The question was left unresolved. Perhaps it was only Columbus's underestimate of the size of the globe that made it seem a present problem at such an early date, for in fact the Oriental limits of Spanish and Portuguese expansion were to remain separated in practice by vast distances almost until the last quarter of the sixteenth century.

On 24 May 1493 jointly with Juan de Fonseca, then Archdeacon of Seville, who was later to become general overseer of Spain's American empire, Columbus was appointed to prepare a fleet in Seville, Cadiz, and other ports. The new expedition was to be much larger and grander than the first, with wider aims, including colonization as well as exploration. On his return from his first voyage Columbus had planned what would essentially be a trading colony in the Genoese tradition, regulating the putative cotton and mastic trades of Hispaniola, the exploitation of gold, and the enslavement and export of the cannibals of other islands. The Arawaks of Hispaniola, meanwhile, were to be evangelized and a group of friars was shipped for the purpose: unhappily, Columbus's record of close sympathy and understanding with Franciscans like Juan Pérez and Antonio de Marchena was not to be sustained with his Franciscan collaborators in the field.

Columbus did not intend, it seems, to settle Spaniards permanently in the Indies, though he expressed a frustrated preference for family men who would behave responsibly and lend stability to the community while they were there; he wanted—but in the nature of things failed to get—skilled craftsmen and industrious prospectors who would stay and apply their skills for a period of years, building up a regular rhythm of production and supply, discovering and developing new sources of trade, supporting by force of arms, if necessary, further exploration and the extension of Spanish control. Gradually, a process of turnover would renew the entire personnel of the factory. How this intention was manifested in his actions and policies, and thwarted in practice, would be the dominant theme of Columbus's life and of the history of his unhappy colony over the next six years. Beyond his colonizing aims—

and, in a sense, in conflict with them because of the incompatible demands on his time and presence—he also had important work of exploration to accomplish, including the reconnaissance of more islands which he so far knew only by report and above all the attempt to verify his belief that Cuba was a promontory of China.

The explorations he engaged in while governing the Indies were to be starred with a fate almost as evil as his colonizing ventures. Yet they began under the influence of success and in the glow of hope. He had mustered an impressive fleet of seventeen sail for the voyage, including the trusty *Niña* which on this occasion was to be under his own command. He was joined by his youngest brother, Giacomo, now called Diego Colón in the Spanish manner, whom Columbus's renown had attracted from Genoa. The total personnel numbered probably over thirteen hundred men, including over two hundred volunteers—the only members of the task force without a salary from the monarchs—and twenty cavalry, whose haughty behaviour and inferior mounts were to affront Columbus. There could be no more emphatic proof of his patrons' confidence or the world's concurrence than the size and superficial grandeur of the expedition. The departure was marked in lavish style, with so much noise of music and gun-salutes, according to one participant, that 'the Nereids and even the Sirens were stupefied'.[12]

Their course from Gomera this time took them sharply to the south of Columbus's former track, so that they made their first landfall at Dominica in the Lesser Antilles on 3 November 1493. On only his second attempt, Columbus had found the shortest and swiftest route across the Atlantic. It was not intended only as a display of navigational virtuosity: Columbus was pursuing an ambition, left unfulfilled at the end of his first voyage, to follow Indian directions to the reputedly rich islands south-east of Hispaniola. A chain of new islands was discovered as he turned north and made for Hispaniola along a route which took him to Puerto Rico, or San Juan Bautista as he named it, through the heart of cannibal country. The first major investigations ashore were made at the island of Guadalupe (now called Guadeloupe), so named in honour of the great shrine of Extremadura which Columbus had visited shortly before his departure from Spain.

If Columbus had once doubted the reality of cannibalism, he was now faced with what his men, at least, interpreted as incontrovertible evidence; the expedition's physician, Diego Álvarez Chanca, who had joined the expedition apparently because he underestimated its discomforts and overestimated his pay,[13] recorded it in a letter to the city council of Seville:

We inquired of the women who were prisoners of the inhabitants what sort of people these islanders were and they answered, 'Caribs'. As soon as they learned that we abhor such kind of people because of their evil practice of eating human flesh, they felt delighted. . . . They told us that the Carib men use them with great cruelty such as would scarcely be believed; and that they eat the children which they bear them, only bringing up those whom they have by their native wives. Such of their male enemies as they can take away alive they bring here to their homes to make a feast of them and those who are killed in battle they eat up after the fighting is over. They declare that the flesh of man is so good to eat that nothing can compare with it in the world; and this is quite evident, for of the human bones we found in the houses, everything that could be gnawed had already been gnawed so that nothing remained but what was too hard to eat. In one of the houses we found a man's neck cooking in a pot. . . . In their wars on the inhabitants of the neighbouring islands these people capture as many of the women as they can, especially those who are young and handsome, and keep them as body servants and concubines; and so great a number do they carry off that in fifty houses we entered no man was found but all were women. Of that large number of captive females more than twenty handsome came away voluntarily with us. When the Caribs take away boys as prisoners of war they remove their organs, fatten them until they grow up and then, when they wish to make a great feast, they kill and eat them, for they say the flesh of women and youngsters is not good to eat. Three boys thus mutilated came fleeing to us when we visited the houses.[14]

Quite apart from these savage habits, the Caribs' warlike aptitude and fierce courage were sources of dismay for Columbus, who began to realize that the conquest of the regions he had discovered would not be as easy as first anticipated. On the other hand, proof of the existence of cannibalism in the Caribbean confirmed the stories told by ancient and medieval writers of anthropophagous peoples in the extreme Orient: it was a misleading hint that Columbus was nearing his goal. So irredeemable a people as the Caribs, moreover, could be enslaved without objection from moralists at home: their offences against natural law put them beyond the pale of its protection.[15]

Full of foreboding induced by their encounters with the cannibals, the Spaniards made their way past Puerto Rico, whose people, Chanca improbably declared, were ignorant of the art of navigation. At last, without hesitation or error on Columbus's part, they arrived off Hispaniola on 22 November. This further proof of Columbus's prowess on the sea left his friend and shipmate, Michele de Cuneo, amazed. 'In my opinion,' he wrote, 'since Genoa was Genoa, there was never born a man so well equipped and expert in the art of navigation as the said lord Admiral.' He went on to single out a type of skill of which Columbus was to give many remarkable proofs in the remainder of his career: 'he had only to see a cloud or a star at night' in order to be able to predict the weather.[16] It was, perhaps, another example of the way his sensibilities, often so deficient in his human relations, were well attuned to nature. It might be a mistake, however, to attribute his course from Dominica to Hispaniola solely to Columbus's intuitive gifts. He carried Indian guides, who knew the waters from their own canoe-borne voyages.

It was an unfamiliar portion of the coast of Hispaniola that confronted them on arrival, but they lost no time in working round to the north towards the fort of Navidad, which had now been without succour, save, as Columbus hoped, from Guacanagarí and the supposedly friendly natives, for more than ten months. Within a week, the fleet was off Navidad, entertaining a party of natives sent out in canoes by Guacanagarí to bid them welcome. The first intimations of disaster struck when the envoys mentioned the outbreak of hostilities on the island; Guacanagarí had been wounded in battle with a rival chief, and the Christians of the garrison of Navidad had all been killed. Columbus was incredulous at the news, but it was gruesomely confirmed by the evidence which the light of the next morning revealed. Navidad had been burned to the ground, and the thirty-nine Spaniards who had remained there had become the first casualties in a long series of colonial wars in the New World. The local Indians' fear that they might be blamed caused them to disperse and hide, thereby only increasing the Spaniards' natural suspicion of them. Columbus was inclined to give them the benefit of the doubt, perhaps attributing the massacre to the work of the Caribs, or accepting Guacanagarí's story of a vengeful attack by a chieftain from eastern Haiti, after the Christians had committed atrocities in the land.

When Guacanagarí's wound proved to be entirely diplomatic, a controversy broke out among the explorers. The faction that demanded vengeance was led, inappropriately, by the missionary leader Fray Bernardo Boil, whose evangelical charity seems sometimes to have been dimmed by natural spite. Las Casas summarizes a lost source from Columbus's own hand:

That Father Boil and all the rest wished to take Guacanagarí prisoner but the Admiral would not do so, though it was within his power, believing that since the Christians were dead the seizure of Guacanagarí could neither restore them to life nor convey them to Paradise if they were not already there, and . . . it seemed to him that the chieftain must be similar to kings among the Christians who have other monarchs related to them, whom such a seizure would offend.[17]

The extract shows that Columbus had a selectively enlightened view towards Indian chiefs, which in general would be maintained both by him and by other servants of the Spanish Crown, who were prepared to ascribe to native polities similarity, albeit not equivalence, with Christian states. It also shows how he had been forced to abandon his original impression of the people of Hispaniola as docile and easily subjected. On the contrary, he was now positively afraid of a native alliance against himself, having seen the destruction wrought in the camp of Navidad.

It soon became clear that the massacred Spaniards had been pursued not only by the natives but by their own sins, anticipating many future colonists with their excesses. Columbus had actually foreseen this danger; his assumption that the garrison would be unmolested had been qualified by the assurance that even if the natives turned hostile, they would remain ineffective.[18] It was no good, however, recalling that feeble warning: the whole thrust of his predictions had been to create a sense of security which events had proved false. The Indians complained that the garrison had quarrelled within itself and its men had gone on a career of woman-snatching and gold-stealing about the island. It would not be long before Columbus was trying to restrain the men of the new expedition from the same depredations. What was perhaps worse, the seeds of the crime the monarchs of Spain were obliged to hate most, the crime of heresy, had been sown on Hispaniola even before the true faith could be preached in earnest, for one of the garrison of Navidad had taught Guacanagarí 'certain things injurious and derogatory to our holy Religion'. Columbus had to 'correct him therein and made him carry a silver image of our Lady about his neck'. In relating such details to Ferdinand and Isabella, Columbus clearly hoped to soften the blow that must fall with news of the massacre and to distract his sponsors from the failure of his own predictions of Indian passivity.[19]

Columbus calmed Guacanagarí's apprehensions and restored good relations between the Spaniards and the locals by presenting him with a large gift of glass beads, knives, scissors, tin bells, pins and needles, and spurs, worth in all four or five *reales* (136–170 maravedis), 'and therewith Guacanagarí believed that he had become very rich'.[20] The task in hand was to choose a site for a permanent township—one of the vessels had been sacrificed to provide the wood and nails—and reconnoitre the country inland, punishing the natives deemed responsible for the Navidad massacre at the same time. Already disappointed in his expectations of the natives, Columbus was becoming rapidly disillusioned with the climate and the land; his men suffered from the unfamiliar environment and diet, and the livestock he had brought from Spain to provision the new colony showed little adaptability. In selecting a location for the settlement, he fell between the pressing demands of speed on the one hand and a salubrious position on the other. He fixed on a foul and ill-watered spot simply because it was to hand.[21] On 2 January 1494, just two years after the commencement of his enterprise, the town of Isabela was founded in a solemn ceremony, the first, and worst-fated, township in the New World.

In reconnoitring and garrisoning the island, Columbus relied heavily on two of his subordinates, Alonso de Hojeda (the Duke of Medinaceli's man and future companion of Vespucci) and Pedro Margarit, who, after attracting from Columbus the usual extremes of friendship and enmity, would soon return to Spain to file complaints against the Admiral. Neither seems to have shared Columbus's cordial enthusiasm for the Indies or to have been motivated by any scientific or evangelical ideals. Like most of their comrades, they were interested in gain and had come to Hispaniola for the sake of its gold, not its people nor land. Columbus established Margarit in a riverside fortress inland, with the aim of recruiting the Indians to work the river-bed for gold, while allowing Hojeda to roam the island in search of mineral wealth and the culprits of the Navidad massacre. Their maltreatment of the natives culminated in the execution of a local chieftain for a theft committed by members of his community—a judicial murder at which Columbus himself connived, partly from a false sense of justice, partly from the pressure of his subordinates, and partly from his own uncertainty about how to treat the Indians. The natives had shown themselves as a potential threat; they were not responding to evangelization as Columbus had hoped; they were proving an inefficient labour force; and by failing to behave as Columbus had predicted, they were undermining his authority with both his patrons and his men. In the circumstances, Columbus was willing to make an example of them. It was another serious miscalculation. The bloodshed increased the belligerence of the 'resistance movement'. While the malleable Guacanagarí remained quiescent, the chieftain blamed for the Navidad disaster, Caonabó, was in the full cry of what the Spaniards called 'rebellion'. When not enjoying the safety of numbers, the explorers were in continual peril.

Deportation and enslavement of the natives on a massive scale, as had been practised in the conquest of the Canary Islands, was the only remedy Columbus could think of, and early in 1494 he determined to begin the shipment of Indians to the Old World. He was impassive before the inherent contradictions of his policy for he was proposing to export the very labour force on which he had planned to rely. He was also breaking the monarchs' explicit instructions for the benign treatment of the natives and was starting on a course of action which canon law condemned for its harmful effects on the evangelization of primitive peoples. Yet political considerations within the colony and economic exigencies—for Hispaniola was still bringing no profit and the amount of gold yielded was minimal—compelled him to exploit the one product that he had ready to hand for export. As the native problem became exacerbated, and the colonists grew restive under the strain of the unhealthy environment and their disappointed hopes in the easeful and auriferous qualities of the soil, Columbus became anxious to leave Hispaniola on a course of resumed exploration by sea. Indeed, life in Isabela seems to have become unbearable. By Columbus's admission, few men were healthy and many—if their own complaints can be believed—were close to starvation. There were reports—or were they just later embellishments?—of ghostly wailings by night and of shadowy processions of headless men, grimly greeting the famished colonists in the streets.[22]

Columbus's own state of mind, and his misgivings about the fledgling colony, emerge from the only document of this period to have survived intact directly from his hand—a memorandum of points to be conveyed to Ferdinand and Isabella by a messenger bound for Spain, written at the end of January 1494.[23] It was an attempt to salvage his reputation from the wreck of his hopes. The structure of the document is revealing. It opens with a cascade of reassurances and praise of Hispaniola's potential: this is Columbus's attempt to bolster royal confidence in his now discredited judgement. A series of damning admissions follow, wrung from the writer with obvious pain. One by one, Columbus's earlier false predictions—about the gold, the climate, the Indians—are stripped away and the horrible reality of life on a savage frontier is exposed. Columbus interweaves excuses with the admissions; some of the disasters—such as the worst of all, the massacre—are reported only obliquely. Caonabó is mentioned only as 'a very bad man and, what is more, a very bold one'. Columbus turns quickly to a vision of the island's future, to which he devotes a great deal of space and detail. Though expressed with enthusiasm and urgency, as a future gage for present failures, the vision is in many ways a bleak one. The islands are to be transformed into ersatz versions of the Old World, planted with wheat, vines, and sugar and grazed by Castilian livestock. The natives are to be subjugated and evangelized, wrenched into a European way of life or exported as slaves. An intruded settler population will engage in petty industry or commerce or the military occupation of the territory; but they will have to consist of more tractable types than those Columbus found among his own men. He pleads for men who will have a stake in the long-term success, rather than just the short-term exploitation of the colony. It was a plea often to be repeated in the history of the Spanish New World. Finally, Columbus turned to his abiding preoccupation with his own share of the profits to be yielded by his discoveries.

He still hoped to shore up one of his illusions by proving the supposed continental nature of the island of Cuba. Having left Margarit in charge of Hispaniola, nominally but ineffectually under the orders of Diego Colón, with a by now inapposite reminder to treat the natives humanely, he put out from Isabela with a portion of the fleet on 24 April 1494 (see Map 4). After interrupting the exploration of the Cuban coastline to make a fruitless search for gold on Jamaica, he began the quest in earnest in the last week of May. Beyond the spiritual strain imposed on him by the frustrations encountered on Hispaniola, the Admiral was now physically exhausted after weeks of taxing navigation amid the shoals and reefs that lie in wait for unsuspecting vessels between Cuba and Jamaica. Whenever he recalled this period, for the rest of his life, the pain came back to the sleepless eyes that 'burst with blood', tortured with watchfulness.[24] And as the days wore on, relieved by little rest and less evidence that they were remotely near Asia, failure began to tell on Columbus's delicately balanced mind.

He took refuge in two characteristic mental bolt-holes: chiliastic fantasy, and the insistence that he had been right all along. He snatched at any evidence, however implausible, that Cuba was part of continental Asia, inventing some of it and reading the rest into garbled associations of native words with place-names mentioned by Marco Polo. He claimed that the footmarks of large animals, including griffins, indicated the Asiatic nature of his discovery. This was not the unfounded assertion it at first appears, since large animals were indeed deemed—by Peter Martyr, for instance, and other savants—to occur only in continental lands. On the other hand, there have never been large quadrupeds on Cuba, and Columbus's visions of griffins can only have been figments of a wishful imagination, stimulated by the febrile effects of frustration, sleeplessness, and overwork. The claim of one of his crew to have seen a white-clad man on the island prompted Columbus to extract from native informants the legend of a 'saintly king of great estate who held infinite provinces and wore a white tunic'. It sounds as though Columbus was imputing to the Indians the image of Prester John, the mythical Christian prince whom some authorities located in Africa and others in the depths of the Orient.[25] At that very same time, a Portuguese emissary, Pedro de Covilhão, was greeting the Negus of Abyssinia under the same name and style.

Finally, that increasingly familiar element in Columbus's syndrome, the obsession with Jerusalem, began to reassert itself, as he talked to his men of leaving the islands to circumnavigate the world and return home to Spain via Calicut and the Holy Sepulchre. The last occasion on which we know Columbus to have reverted to talk of his Jerusalem project was towards the end of his first voyage, when he was anxiously contemplating the consequences of the loss of the *Santa María*, and turned to the hope that the garrison of Hispaniola would gather so much gold 'that within three years the monarchs would undertake and prepare to go to conquer the Holy House'. 'If I ever forget the thought of thee, Oh Jerusalem, let my tongue cleave to the roof of my mouth!' For Columbus, the thought of Jerusalem seems to have acted like a form of penance, when his conscience was uneasy and his confidence low.[26]

After more than three weeks of dangerous sailing along the Cuban coast, fed by frustration and fantasy, Columbus decided to abandon the exploration of Cuba. He convinced himself that he had explored 370 leagues of coast—an egregious overestimate—and claimed that no island could be so large. On that basis, he called on the ship's scrivener, who combined the functions of scribe and public notary, to record the oath of almost every man in the fleet that Cuba was a mainland and that no island of such magnitude had ever been known. The declaration was false on both counts, but Columbus had been taxed so much by his grim experiences that he was now beyond the influence of reason, and the men made little attempt to argue with him. They further swore that had they navigated farther they would have encountered the Chinese— obviously a rash claim to make on oath—and promised to abide by the opinion to which they had sworn on pain of a fine of ten thousand maravedis and the loss by excision of their tongues.[27] To exact such an oath and threaten so brutal a punishment were not the acts of a man exercising rational self-control. The only point that can be made in Columbus's defence, apart from the extenuating circumstances, is that he may have been the victim of a computational error of his own. For he had taken the opportunity of an eclipse, while in western Hispaniola, to attempt to calculate his longitude. He made the time difference from Cadiz, where the table of the predicted times of eclipses he had with him was presumably set, a matter of ten hours, corresponding to 150 degrees of longitude, which, to a small-world theorist, might reasonably have signified the neighbourhood of China. Columbus shared the examination-candidate's curse: the method right, the answer wildly, alarmingly wrong.[28]

What is more, Columbus seems to have known in his heart that he was perpetrating a lie. His friend from Savona, Michele de Cuneo, was pardoned the oath, and the penalty clause shows how unsure Columbus was of the support of his men in his claims. Most of the crew probably took the oath merely to placate the Admiral, or out of fear that he might try to execute his threat of circumnavigating the globe via Jerusalem. At home, no one seems to have taken the myth of a continental Cuba seriously for very long, although a puzzling promontory shown in the earliest cartographical tradition of the New World (that of the Cantino map of 1502) may represent an attempt to show both an insular and a mainland Cuba.[29]

Little solace in his distress awaited Columbus in Hispaniola. He arrived towards the end of June, to be greeted by his brother Bartolomé Colón, who had at last returned after five or six years in France and England. There was some comfort in seeing Bartolomé, whom Columbus made his deputy with the title of *Adelantado* of the Indies, but none in the news he brought. Complaints against the beginning Columbus had made in the government of Hispaniola had already reached the monarch's ears. In particular, Fray Boil had never established good relations with the Admiral from the moment of their first disagreement over the fate of Guacanagarí, and had returned to Castile to impugn Columbus. None of the colonists was satisfied with the state of affairs on Hispaniola and many had taken the opportunity to relay their complaints with the home-bound fleet that had left the previous February. Some charges can be inferred from surviving sources: with the next fleet, for instance, Columbus had to send documents exculpating himself from the charge of embezzling royal gold.[30] An anonymous memorandum of 1496, by a critic who evidently knew Hispaniola at first hand, pointed out how little gold there was, in relation to Columbus's promises, how the Indians were incapable of supplying the quantities demanded from them, how the so-called spices of Hispaniola were worthless, and how cotton was the most promising of the potential sources of trade.[31] It is tempting to assume that the friars opposed the policy of unlawful enslavement of Indians: certainly, they complained that Columbus's conduct was an obstacle to evangelization. Ferdinand and Isabella regarded Columbus's practice as both a crime and a mistake. Their objections were partly juridical, for they doubted the propriety of enslaving potential Christians, and partly religious, for it was a commonplace that converts had to be won 'with love'; but they were also swayed by political considerations: while the Indians remained free, they were under the direct dominion of the Crown, whereas enslavement would transfer them from the immediate lordship of the monarchs into the control of private proprietors. The trade, in any case, was impracticable as well as unlawful, since the slaves died in droves on the voyage to Spain or perished unacclimatized on arrival. Indians unlawfully enslaved by Columbus were freed in Spain after judicial review, and ordered to be shipped home.[32]

The main allegation against the Admiral's rule is apparent from the defensive tone of the communications home of Columbus and some of his friends. He had broken faith with his men by misleading them about the attractions of the island, and indirectly caused the sickness, and in some cases death, to which the hostile environment condemned them. His efforts to deflect this charge were unimpressive. He was, perhaps, right to attribute the prevailing malady to the 'airs and waters' of their new home, rather than the syphilis which later writers have alleged. Dr Chanca's diagnosis, which also blamed the anaemic diet, deficient in red meat and wine, bore him out.[33] Yet these excuses missed the point. The main significance of the epidemic was the lie it gave to Columbus's glib promotion of an ideal island and a holiday-climate.

If the news from home was bad, the situation in Isabela and throughout Hispaniola on Columbus's return was worse. The colonists were openly at war with the natives. Rebellion against Columbus's authority was brewing among the embittered subordinates; the crops brought from Europe were performing badly in the unfamiliar soil and provisions from Spain had been exhausted by the spring of 1494. Gold production dwindled as native artefacts ran out; extraction was still at a modest level, amateurishly undertaken.

Columbus, who in Cuba had seemed on the verge of a total personal breakdown, responded to these problems with the sort of resilience and energy which he often showed in a crisis. His overriding objective was to silence his detractors and staunch the flow of calumnies and complaints which, if unchecked, would alienate the monarchs' precious favour and jeopardize all he had achieved. The result was that short-term expedients took the place of long-term planning and the interests of the natives were forgotten in the desperate effort to appease the colonists. Apart from continuing with the disastrous policy of enslavement, Columbus adopted two methods of dealing with the Indian problem: vigorous punitive action, and the exaction of tribute, backed by the building of a chain of forts. This policy doomed the Indians, who had never faced laborious work or heavy taxation before, to hardship, despair, and in many cases early death under the strain. Columbus, who could not anticipate the impact of the 'cultural shock' of imposing Old-World values in America, was guilty of misjudgement rather than wickedness, misjudgement the worse in that tribute could at best be effective only as a short-term remedy. The natives could be made to hand over the golden trinkets they had accumulated over many years of slow extraction of impure ores in recondite places, but until a systematic exploitation of the island's gold was devised, there was no prospect of maintaining the flow of gold into Spanish hands. As things were, once the Indians' long-standing stocks were exhausted there would be nothing to replace them with. And the cotton with which Columbus proposed to make up any deficit in default of gold was a substitute that could satisfy no one. In defence of the policy it must be said that some tribute had to be exacted as an affirmative sign of the vassaldom which, by the Pope's command, the Indians now owed to the Spanish monarchs; and since the colony was an expensive undertaking for the Crown, it was essential to provide some quick return.

The punitive measures adopted by Columbus were vigorous and wide-ranging enough to justify his later boast to have 'conquered' Hispaniola. A series of campaigns from late 1494 onwards took him, Hojeda and Bartolomé Colón to almost every corner of the island. Columbus expected great consequences from the capture of Caonabó, which Hojeda effected by a ruse at a sham parley, reportedly persuading the chief to don 'bracelets' which were really manacles. Indian restiveness and resistence, however, were at best unaffected and perhaps even stimulated as a result. The 'good harvest', as Las Casas ironically called the war, served only to disrupt production and deplete manpower. Las Casas's claim that it wiped out two-thirds of the population is no doubt unreliable—but an event must be terrible enough in reality to acquire such proportions in legend. Peter Martyr, whose view of the natives was more detached than that of Las Casas, reckoned the dead at fifty thousand, blaming famine induced by the Indians' scorched-earth tactics. Gonzalo Fernández de Oviedo, who shared most settlers' prejudices against the Indians and blamed the outbreak of violence on their unwillingness to collaborate with the invaders, spoke of victims without number.[34]

While the Indians or Columbus's subordinates—particularly Alonso de Hojeda—tended to get the blame for starting the bloodshed, Columbus, in the same sixteenth-century sources, was generally credited with ending it. At the end of March 1495 he led an impressive column into the heart of the island: 200 Spanish foot, twenty horses and twenty dogs, with a force of native auxiliaries under Guacanagarí, scattered the insurgents 'like flocks of birds'. In the centre of Hispaniola, Columbus built a new fort, known as Concepción de la Vega, and received acts of submission and promises of tribute from 'many chiefs'. According to the Columbus legend the island was brought in consequence 'to such peace and security that a Christian could go anywhere in safety alone, and the Indians would carry him wherever he wished on their shoulders like pack-horses'. In Oviedo's time, the place was a shrine with a reputation for sanctity, and miracles were worked at the foot of a cross Columbus erected.[35] In reality, the peace was an illusion. Most of the tributed was never delivered and Bartolomé Colón had to undertake another bloody campaign the following year. It is remarkable, none the less, that Columbus was always to look back with pride on his achievements at Concepción. In his last will, he expressed a wish to endow a chapel for his soul and those of members of his family, in Hispaniola, 'which God gave me by a miracle and it would please me if it were in the place where I called upon Him, which is called Concepción de la Vega'.[36]

It was not to redress the plight of the Indians, but rather the complaints of the colonists, that Ferdinand and Isabella commissioned the first judicial inquiry into Columbus's discharge of his offices in 1495. This was a regular procedure, but it was unusual for it to happen during the incumbency of the official who was the object of the inquiry.[37] The inspector, Juan Aguado, arrived in October 1495. He and Columbus knew each other well. His was another example of Columbus's inability to keep a collaborator as a friend. As a member of the expedition that made the second crossing of the Atlantic, he attracted commendation from the commander for 'having served well and faithfully'. But some unknown revulsion from Columbus's cause sent him back to Spain, presumably with Boil and Margarit. He was thus deeply tainted, in Columbus's eyes, as a member of a faction of enemies. What he unearthed now against his former master, when he returned with every opportunity to spite him, is unknown. Columbus's own subsequent references to the episode imply satisfaction with its outcome. The experience seems to have convinced him that the struggle for authority in his colony could only be won at court, and he prepared to withdraw to Spain, perhaps specifically to plead in connection with Aguado's report, early in 1496. Considered from one perspective, his departure looks like a flight from intractable problems and uncongenial circumstances. Yet he had recovered from the worst crisis in his fortunes: he had restored uneasy peace to the colony; he had sold the Indians and bought the colonists; he had retrieved his old address after the disaster of the Cuba mission. He had emerged unbowed from Aguado's investigation. He could hand over his authority to Bartolomé and leave with some hopes of better success to come. The fair crossing in the *Niña* in March 1496 may have helped to clear his head, beset still by thoughts of the Amazons. But the recovery of his health, as of his fortunes, was only temporary. While he accumulated instructions from Ferdinand and Isabella over the next two years for the future administration of his colony, planning his return, fending off his calumniators, and dreaming up new projects of exploration, he was dangerously unaware of the worse trials and more damaging failures that were still to come.

6

'Your Will to Continue in this Enterprise'

JUNE 1496–AUGUST 1498 AND THE THIRD CROSSING

THOUGH his reputation survived the investigation by Aguado, Columbus's reception home from his second voyage was attended by none of the encouraging junketings, none of the intoxicating adulation which he had enjoyed in 1493. The disaffection of the colonists and the distancing of Asia emboldened his critics and alarmed his patrons. The next two years of Columbus's life, before he was granted leave for a third ocean crossing, were dominated by anxiety about the hard-won concessions he had wrung from the monarchs. As he summarized the bitter debate in which he found himself involved:

Evil words arose in Spain and belittlement of the enterprise which had been begun there, because I had not at once dispatched ships laden with gold, without allowance for the brief time that had elapsed nor all else I had said of the many problems. And in this business—I believe it must have been for my sins or my salvation—I was placed in abhorrence and obstacles erected against whatever I said or asked. Therefore I decided to come before your Highnesses and express my incredulity at all this, and show you how I was right in everything. And I told you of the peoples I had seen whose many souls—some or all of them—could be saved. And I brought you the submissions of the people of the island of Hispaniola . . . and a sufficient sample of gold . . . and . . . of many kinds of spicery . . . and an infinity of things beside. Not all this was good enough for some people who had a fancy to speak ill of the enterprise. . . . Nor did it avail to point out the things which great princes customarily did in this world to increase their fame, such as Solomon, who sent from Jerusalem to the ends of the Orient to see the mountain of Ophir, where his ships tarried for three years and which your Highnesses now possess in the island of Hispaniola; or Alexander who sent out an expedition to see the government of Taprobana in the Indies; or the emperor Nero to the sources of the Nile. . . . Nor did it avail to say that I had never read that princes of Castile had ever before won territory outside Spain, and that these lands are another world which the Romans and Alexander and the Greeks had striven to conquer, with great exercise of arms; nor to point out the present achievements of the Kings of Portugal, who had the heart to persevere in Guinea and in the discovery thereof . . . The more I said, the more these calumnies they uttered were redoubled and abhorrence shown. . . . Your Highnesses answered me, smiling and saying that I should not be anxious about anything, for you gave no heed or credence to those who spoke ill to you of this enterprise.[1]

The confidence he expressed in his patrons sounds like bravado. He might have been pardoned for thinking that the monarchs were keeping him in enforced inactivity, delaying or denying his return to the New World. He showed unmistakable signs of brooding: by the time he was to leave once more, in the spring of 1498, he had evidently spent a great deal of leisure systematizing the supposed references to his discoveries in classical and biblical sources, on which he drew in his arguments to the monarchs. He arrived at the conclusion that God 'spoke so clearly of these lands through the mouth of His prophet Isaiah in so many places in His scriptures, announcing that His Holy name would be spread abroad from Spain'.[2] This remarkable insight may have been based, in particular, on Isaiah 60: 9, which reads, in the Jerusalem version:

> Why, the coasts and islands put their hopes in me
> And the vessels of Tarshish take the lead
> In bringing your children from far away
> And their silver and gold with them
> For the sake of the name of Yahweh your God.

Tarshish was commonly and plausibly identified as Spain. Las Casas endorsed Columbus's view that Isaiah had foretold 'that from Spain would come the first men to convert these peoples' but conveniently regarded it as presumptuous to try to single out particular texts.[3] The habit of ferreting for scriptural prophecies of his own work became first a habit and then later—after Columbus's third voyage—an obsession which fed the providential and messianic delusions that would come to grip him in later life.

He found some practical distractions, during 1496 and 1497, in plying the monarchs with memoranda about the government of the Indies: about the numbers of colonists for Hispaniola, the siting of settlements, the administration of municipalities, the provision of a divine ministry, the licensing of mining, the encouragement of farming, the surveillance of trade, the disposal of the estates of deceased colonists. He made some investments in the shipping of foodstuffs to the colony, and began negotiations with Genoese bankers in Seville for capital for a third crossing.[4] In the summer of 1497 he spent some time in retreat in the Franciscan house of La Mejorada, where—as he later recalled—he wrote proposals for a crusade against Mecca and a spice-buying voyage to Calicut—both, presumably, by way of the west.[5] Calicut was Vasco da Gama's destination on the great voyage then being prepared in Portugal, and Columbus's thoughts may have been stimulated by an emulous purpose.

The doubts that clouded the monarchs' view of their discoverer were dispelled or, at least, temporarily lifted by early 1498. By February of that year he was actively preparing his departure on a venture that was both to extend the colonization of Hispaniola and increase the range of the exploration of the Indies. Ostensibly, he had retrieved the full confidence of the monarchs but in painfully evident reality, their favour was now provisional—conditional upon some solid success. The most curious of the documents he drew up before leaving is the entail of his estate, made in Seville on 22 February 1498.[6] The monarchs' willing-ness to allow him to draw up an entail was a very marked sign of their renewed favour, conceded—as a rule—only to aristocratic families whom they wished particularly to favour, and whose wealth they were willing to see conserved in dynastic hands. It was therefore a celebration of the new role Columbus relished as a Castilian nobleman. Such documents were normally drawn up by notaries in accordance with well-tried formulae. Columbus's, however, while influenced by professional help, bears the hallmark of his personal literary creation. Indeed, it appears from some linguistic slips to have been dictated to the notary. Some of its provisions are bizarre, and much of its language unprofes-sional. It is prolix and repetitive, even by the formidable standards of Spanish legal discourse of the time. Yet, like much of Columbus's writing, it retains enormous natural power and vivid, personal force. Seven themes emerge more or less strongly.

First and foremost is the obsession with lineage. The document in which he created his entail was for Columbus the act of foundation of an aristocratic dynasty. Succession would be limited, in normal circum-stances, to the nearest legitimate male heir and to those who 'call themselves and shall always have called themselves' by the name of

Colón. Columbus stressed that the procedures he enjoined were normal 'for persons of title'. He repeatedly envisaged the transmission of the entail 'in perpetuity' and 'from generation to generation'. He explicitly compared himself with the hereditary Admiral of Castile. Egregious social ambition, the great driving force of his life, had brought him from the weaver's loom and continued to colour his vision of the future.

The second theme is constituted by his references to the terms of his bargain with the monarchs. Obviously, this was highly germane to his purpose: it was the basis of his claim to heritable titles of nobility and to the material rewards on which he hoped to establish the greatness of his house. Columbus's insistence, however, seems to have gone beyond what the occasion required and to betray his anxiety over his prospects of exacting performance of his contract with the Crown: hence in part, perhaps, his dwelling on the extent of his discoveries, which was evidence that he had fulfilled his side of the bargain with his patrons, and the hint of divine authority for his claim to have discovered the Indies—'it pleased our Lord Almighty . . . Our Lord gave me victory . . . The Holy Trinity put into my mind the thought, which later became perfect knowledge, that I could sail to the Indies from Spain by crossing the Ocean Sea to the west.' Columbus's prospects of obtaining in practice the rewards to which he staked a claim were rendered more problematical by his extraordinarily generous estimate of what was due to him: 25 per cent of all the yield of the New World. Ferdinand and Isabella were never prepared to admit that they had conceded more than a tenth part of their own share of one-fifth of whatever might be obtained that was subject to a royal levy.

Though this in itself was enough to make any individual or family wealthy, the amounts of money with which Columbus juggled in his imagination were fantastic. Unrealistic pecuniary ambitions are the third great theme of the document. Columbus envisaged fortunes of millions being piled up by collateral branches of his family, the dowries of poor relations being provided, charitable endowments accruing with a lavish hand and money accumulating for the ever-promised campaign to liberate Jerusalem. This esoteric reference to the millennial city should perhaps be considered in conjunction with the fourth theme—that of Columbus's cryptic signature, which all his direct heirs were ordered to use. The device first occurred in a document of 1494 and had gradually come to displace other forms of Columbus's signature. It was arranged in three columns and three lines: the first line contained a capital S, flanked by dots, in the central column; the second an 'S', 'A', and 'S', one in each column, divided by dots; and the third an 'X', 'M', and Greek 'Y', similarly distributed, but without dots. Under it, his heirs are instructed to write 'el Almirante' ('the Admiral'), with no further elucidation. The meaning of the strange device is left pointedly unexplained. Alain Milhou has recently argued that the arrangement of symbols is meant to correspond to images in the traditional iconography of the Coronation of the Virgin, with each 'S' representing one Person of the Trinity ('Sanctus, Sanctus, Sanctus' in liturgical allusion), grouped around the Virgin's crown, and the 'X' and 'Y' standing for St Christopher and St John the Baptist respectively—both, like Mary herself, 'bearers' of or for Christ for the world. Christopher, Columbus's own name-patron, bore Christ on his shoulders, Mary bore Him in her womb, and the Baptist bore Him, in the form of the Divine Logos, in his mouth, in the words he uttered to prepare His way: this last role was analogous to that which Columbus later ascribed to himself as the 'messenger of a new heaven', and harbinger of the gospel in a new world.[7] Columbus, punning on his Christian name, commonly called himself 'Christo ferens'—'bearer for Christ'; nor would it have been foreign to his nature to imagine himself in saintly company. If this interpretation is correct, it does not necessarily exclude other possible readings of the signature, which may well have been designed to be intelligible at a variety of levels.[8] Columbus dedicated the voyage he was to make in May 1498 to the Holy Trinity. Together with the aggressively austere manners he adopted at the time, his effort to perpetuate the use of the mystic signature is strong evidence of his growing revulsion, through disillusionment, from worldly standards of success. His interests were shifting towards the evangelical possibilities opened up by his discoveries and towards the potential significance of his own role as a providentially designated child of prophecy.

This reflection has, however, to be set against the limited vision of the future of the Indies revealed in the document creating an entail. It was not, of course, an appropriate context for a detailed exposition of Columbus's plans for the colony he had founded. It is remarkable, however, that only three small religious foundations were specifically envisaged in the document; that these included substantial provision for members of the Columbus dynasty, and only very modest provision for putting the evangelization of the New World on a sound doctrinal footing; that the overall provision was niggardly by comparison with the amounts bequeathed for the glorification of Columbus's descendants; and that one of the major purposes to which Columbus's prospective foundation of the church of Santa María de la Concepción was to be consecrated was the display of the terms of the entail as a perpetual memorial—and, by implication, a perpetual admonition to his heirs. Moreover, the document provided for the diversion of tithes to the enrichment of Columbus's brother, Bartolomé, and to his heirs, pending the accumulation of a substantial fortune. Columbus's piety, if it was felt in his heart, was evidently not meant to be felt in his pocket. The paradox has never been noticed and may never be resolved. Perhaps practical charity or churches of stone are easily forgotten or overlooked by those whose Christianity is of a strongly mystical flavour.

Finally, the entail is concerned with two themes which might have been linked in Columbus's mind: pride in his Genoese origins and implicit dissatisfaction with his Spanish sovereigns. The first is shown by Columbus's repeated asseverations of his Genoese birth, his praise of Genoa and of the Genoese State Bank, and his desire to maintain a house, at the expense of his estate, in Genoa for ever. Columbus's professed expectation that Genoa would help his lineage in the future may have been intended to have, for the monarchs, a threatening resonance in conjunction with his assurances that he came from Genoa to serve them. The implication is that the trajectory could be reversed and his services returned, if inadequately appreciated in Spain, to the city of his birth. The Admiral's reproach to the monarchs for their delay in adopting his plans reveals a note of bitterness which would come to dominate his attitude to his patrons: 'under almighty God it was their Highnesses who gave me the means and right to conquer and achieve this entailed estate, even though I came to these realms to offer this enterprise to them and they spent a long time without giving me the means to put the work into effect.' His insistence that 'they have gone on granting me favours and much increase' is transparently insincere: Columbus is not offering thanks for favours received but trawling for more. Attempts to heap coals of fire on the monarchs' heads, particularly by thanking them for unfulfilled promises, would become a dominant technique of Columbus's supplications to Ferdinand and Isabella.

His two years in Spain also gave Columbus time to improve his command of cosmographical tradition, and of the latest additions to it. He sent to England for information about John Cabot's crossing of the North Atlantic in 1496 and for books and maps to supplement his reading.[9] He seems also to have gone back over old favourites, including Pierre d'Ailly, Pliny, Pius II, and Marco Polo: his references to them after this date are more assured, sharper and more systematically deployed than before—albeit, in sum, no more convincing. The need to defend, against the savants' arguments, his theories of the smallness of the world and of the accessibility of Asia continued to absorb a good deal of energy. He seems to have refined his own arguments, and to have marshalled a copious if disparate array of texts in his support. The defence he wrote towards the end of 1498, when he was back in Hispaniola, demonstrates both the increased stature and the continuing limitations of his erudition as a geographer. D'Ailly's references—to a scholar, embarrassingly unverified—continue to supply the bedrock of his case; disparate authorities are lumped together without sorting, like compacted waste; thirteenth-century theologians jostle classical and Arabic philosophers; fathers of the Church are given equal weight with a Roman playwright and an apocryphal prophet. Like Chinese whispers, the authorities amassed are garbled on their way. The opinion attributed to Aristotle, for instance, by d'Ailly, and appropriated by Columbus— that 'the sea is small between the western extremity of Spain and the eastern part of India'—is unconfirmed in any genuine surviving text of Aristotle's, whose known view seems rather to suggest the opposite.[10] Some delicious irrelevancies are indulged in to add to the species of learning:

Pliny writes that all the sea and land together form a sphere, and he states that this Ocean Sea constitutes the major body of water and that it is placed towards heaven, and that the land is below it and upholds it, and that the two are combined like the meat of a walnut with the thick covering which goes around it.[11]

Was Columbus himself convinced by what, with due respect for effort, was rather a farrago of amateurism? His comfort in the face of learned criticism does not seem to have increased after this time, to judge from the acerbity with which he denounced the savants. Even when plying his readers with copious allusions to the received tradition, he continued to emphasize that his own claims to superior wisdom were based on experience, not book-learning. The decisive argument in favour of a small world was, he said, that it had been demonstrated empirically—by implication, by himself—and he added a line of proverbial lore to drive the point home: 'as for this matter of the span of the earth, it has been demonstrated experimentally that it is very far removed from what is commonly supposed. Nor is this to be wondered at, for "as one goes, one's knowledge grows".'[12]

Columbus left in the last week of May 1498, bound—it might have been supposed—for an urgent appointment in Hispaniola with Bartolomé Colón. Bad news had continued to accumulate from the island. The intractable Indians, the recalcitrant colonists, the implacable mosquitoes, the insalubrious 'airs' and 'waters' that Columbus had left behind at his last escape were still plaguing the colony. A letter Columbus wrote to his brother, some three months before embarking, shows that he had a lively sense of Bartolomé's predicament. Bemoaning the problems of accounting for the gold shipments, the Admiral broke into a poignant *cri de cœur*:

Our Lord knows how much anxiety I have suffered wondering how you are. So these problems, though I may seem to make heavy weather of them, have been far worse in reality: so much so that they made me weary of life because of the great trouble I knew you must be in, in which you should think of me as united with you. Because although, to be sure, I have been away over here, I left and keep my heart over there, without a thought for any other thing, constantly, as our Lord is my witness; nor do I believe that you will have any doubt of it in your heart. For besides our ties of blood and great love, the effects of fortune and the nature of danger and hardship in places far removed embolden and oblige man's spirit and sense to endure any trouble that can be imagined—there or in any other place. It would be a thing of great advantage if this suffering were to be endured for a cause which redounded to the service of our Lord, for whom we ought to labour with a joyful mind. Nor would it be other than a help to remember that no great deed can be accomplished except with pain. Again, it is some consolation to believe that whatever is achieved laboriously is treasured and esteemed as sweeter for it. Much could be said to the purpose, but as this is not the first cause for which you have suffered or which I have seen, I shall wait to speak of it with more time to spare, and by word of mouth.[13]

Is it possible to read these lines without conceding an assumption—if not embracing a conviction—of the writer's sincerity? Or to suppose other than that he would sail directly to his brother's aid at the first opportunity? Yet Columbus departed on his third ocean crossing determined to delay his return to Hispaniola. He seemed gripped by the same evasion, the same distaste, as had driven him prematurely from the island in 1494, the same preference for exploration over administration that fatally compromised his chances of success as the viceroy of a fragile raj.

He divided his fleet into two squadrons. One, of five ships,[14] for the succour of the colony, was to sail directly by the route he had established on his second crossing; another, under his own command, was to make a huge exploratory diversion into an unknown area of the Atlantic (see Map 3). He expected to improve his chances of finding precious commodities by taking a more southerly course than on his previous voyages. It was widely believed in his day that lands on the same latitude would have similar products. It was also thought that these latitudinal zones or 'climates' grew richer the farther south one went. This was the burden of advice the Admiral received from Jaume Ferrer.[15] Columbus therefore decided to drop to the parallel of Sierra Leone, where the Portuguese had discovered gold in the land of the Blacks, and to make his westing from there. He had it on Ferrer's authority that 'turning at the equator . . . where the natives were Black or dark' there also would he find an abundance of precious things. The good relations subsisting then between Castile and Portugal enabled him for the first time to frequent in safety latitudes where the Portuguese held sway.

There was probably a further consideration which influenced his choice of route. According to his own record of the voyage, he was concerned to test a theory he attributed to King João II of Portugal: that an unknown continent existed 'in the south'.[16] No more detailed information about the opinion of the Portuguese monarch exists than what Columbus reports. It is uncertain what the sources or exact nature of the theory were, or even precisely where this land was to be located. It may have been an echo of Macrobius's rumoured southern continent, or a re-interpretation of the legend of Antillia, or a theory, such as Peter Martyr espoused, of the nature of the discoveries Columbus had already made; or a version of the legend of an antipodean land known as the Hesperides, which are depicted on some fifteenth-century maps.[17] It may have been connected with rumours, recurrent since at least 1448, that a new land had been sighted deep in the ocean to the west of the Cape Verde group.[18] Or perhaps it was a deliberate fiction, strewn in the wind like a siren's call by the wily King João, to lure Columbus off course. Certainly it was the beginning of a persistent tradition and the 'Unknown Land in the South' would adorn innumerable maps, and inspire innumerable voyages, until Captain Cook finally demonstrated that, if it existed at all, it must be uselessly remote.

In Columbus's mind, the Portuguese rumour raised the suggestion that an unknown continent might lie on his course along the middle latitudes of the Atlantic. During his third crossing it contributed to a chronic scruple that the region in which his own discoveries were situated might enclose such a land: antipodean rather than Asiatic, austral rather than Oriental. His references to the King's theory are

important for an understanding of the problem of Columbus's awareness of the nature of his discoveries, for they prove he anticipated an
encounter with just such a new continent as America turned out to be.
His mood, anticipating new revelations with the same excitement as
preceded his first voyage, is captured in the instructions he wrote for the
supply vessels sent ahead to Hispaniola: 'May our Lord guide me and
lead me to something that may be of service to Him and to the King and
Queen, our lord and lady, and to the honour of Christendom. For I
believe that this way has never been travelled before by anyone and that
this sea is utterly unknown.'[19]

At first the experiment bade fair to succeed. After victualling as usual in
the Canaries, Columbus made a fair passage to his new point of
departure, the Cape Verde Islands. Here was a wilder frontier of
Christendom even than Hispaniola itself. The tone of life there comes
through the will of the founding Captain-General of the islands, Álvaro
da Caminha, written a few months after Columbus's visit, bestowing the
copper utensils, slaves, and sugar futures which constituted his wealth,
expressing anxiety over the paucity and penury of the colony, the
indifference of the metropolis, the menace of runaway slaves. His
nephew and successor, Pedro Álvares, dreamed of building a city 'which
after it is finished will be one of the most magnificent works one could
find'; but it was a hollow dream: the permanent colonists numbered only
fifty, nearly all of them exiled criminals; the shortage of food made it
impossible to accommodate more. The land was 'evil' and there was no
truck to trade for the riches of the mainland. Only Santiago had a
settlement. Most islands were deserted save for lepers seeking the
reputedly prophylactic turtles' blood, or Castilian dye-gatherers or
merchants collecting the shells that could be traded on the African
mainland. On the island of Boa Vista, Columbus victualled with the
meat of wild goats, abandoned by a failed settlement. The effect of the
oppressive and primitive islands blighted Columbus's mood: perhaps
they were too suggestive of what Hispaniola might become. Their name,
he said, was misleading 'for they are all so dry that I saw no green thing.
With all my men sickening, I dared not tarry.'[20]
    A few days out from Santiago, he sailed into the doldrums—the
windless, marine no man's land between the zones of the north-east and
south-east trades—to find himself becalmed, in mid-July under a
savage sun. The heat turned the ship's wine to vinegar, the water to
vapour, and the wheat to ashes; the bacon roasted or putrefied. 'And all

at once everything fell into disorder for there was no man who would dare to go below deck to mend the casks or see to the stores.'[21] But for the overcast skies that came to their rescue during much of the eight days spent becalmed, they could not have hoped to survive. Columbus's account of their escape from the doldrums is confused. On the one hand, he claims that he had reached his intended latitude—that of Sierra Leone; on the other, he makes it plain that he would have gone further south had the wind permitted. As it was, the most urgent priority was to evade the mephitic heat:

I recalled that in sailing to the Indies I found that each time I passed one hundred leagues to the west of the Azores, the environment changed for the better all over, in both the more northerly and the more southerly latitudes. And I decided, if it should please our Lord to grant me wind and fair weather to enable me to get away from where I was, that I would no longer attempt to go further south nor yet turn back, but sail westward until I reached that line, in the hope of finding there the same easing of conditions that I had found when sailing along the parallel of Gran Canaria; and that if it should so prove, I should then be able to head further south.[22]

Potentially, it was a disastrous strategy, condemning the expedition to cling to the doldrums. But a lucky south-easterly, unwonted in that season, rescued the fleet from its ordeal and bore it towards the west. By the end of July Columbus suspected that he was approaching the meridian of Hispaniola but he had not yet noticed any indications of new land. He was well beyond the demarcation line that separated the zones of Castilian and Portuguese expansion, and was at least satisfied that there was nothing on the parallel he had sailed that lay on the Portuguese side. The southern continent, if it existed, appeared to have eluded him. He now took an uncharacteristically pusillanimous decision: while the wind was still favourable he would victual and take on water in the Lesser Antilles, which he had visited on his previous voyage and which he rightly estimated to lie north of his position. When he changed course to the northward, he was unaware that the continent of America lay but a short sail to the west, roughly at the point just south of the Orinoco delta, where the coast of modern Venezuela turns south towards Brazil.

As it happened, fate did not cheat him of his most spectacular discovery so far, for he made the mainland of the New World a few days later; but his change of course dismissed the expectation of a new continent in the offing, which had been preying on his mind at the start of the voyage, and turned his thoughts to the islands he anticipated on his new track. Thus when he stumbled on America, his confusion about its nature—whether insular or continental—was all the greater.

In the period of this voyage, Columbus was going through a phase of particularly strong devotion to the Holy Trinity, whom he evoked at every opportunity, and to whom he had specifically dedicated this third crossing of the ocean. When, on the last day of July, he sighted land for the first time on this voyage, in the form of three low but distinct hills just visible to the north-west, he was pardonably struck by the potency of the coincidence, in token of which, as 'we said the Salve and other canticles and all of us gave many thanks to the Lord', he named the island Trinidad:

And it pleased our Lord that thanks to His divine Majesty the first sighting was of three hummocks, or I should say three mountains, all seen at once in a single glance. . . . For it is certain that the discovery of this land, on this voyage, was a great miracle, as much as the discovery made on the first voyage.[23]

The unity of the glance in which the trinity of humps was revealed was a nicely calculated piece of theological semiotics. The island was fertile and well-watered enough to redress the losses of supplies to the torrid weather of the voyage. To Columbus's eye, it also reawakened his hopes that he was in the Orient: the inhabitants resembled neither the Blacks who lived in the same latitude on the other side of the ocean, nor the Caribs and Arawaks familiar further north. Rather they looked 'like Moors', wore 'turbans' and displayed commercial sagacity. As usual, however, Columbus's first impressions were misleading: the turbans were merely coloured cotton headbands and the commercial nous a prelude to hostility.

As he reconnoitred the coast, it was the curious sea conditions that most impressed him, indeed frightened him when he approached the channel between Trinidad and the mainland—the Tierra de Gracia, Columbus called it—where the Orinoco debouches into the sea. He recalled it two or three months later,

When I arrived at the Punta del Arenal, I observed that the island of Trinidad forms a large bight, two leagues broad from west to east, with a land I called the Tierra de Gracia, and that in order to get into it to complete the circuit of the island to the north, there were some currents to take into account, which crossed that bight and made a very loud crashing noise, like a wave that goes and breaks and dashes against rocks. I dropped anchor off the aforesaid Punta del Arenal, outside the said bight, and I found that the current was flowing from east to west with all the fury of the Guadalquivir in flood. And it went on continuously, day and night, so that I thought I should be unable to get back, because of the current, or go forward, because of the reefs. And during the night, when it was already very late, and I was up on deck on my ship, I heard a terrible roar which was approaching the ship from the south, and I stopped to look, and I saw the sea arisen from west to east, like a broad hill as high as the ship. And still it came on towards me, little by little, and on top of it I could see the trend of the current, and on it came roaring with a mighty crash, like the fury of the crashing of those other currents which, as I said, seemed to me to be like the waves of the sea that dashed against rocks. For to this day I can feel the fear in my body that I felt lest they should capsize the ship when they got underneath her. And it passed by and reached the mouth of the bight where it seemed to teeter for a long time.[24]

No estuary like that of the Orinoco, discharging so vast a volume of water with such great force, had ever been observed by a European before, and Columbus revolved for a long time the bafflement it caused in his mind. Towards the end of the first week in August, a 'nimble caravel' he had sent into the bight to reconnoitre confirmed the presence of a great river 'and everywhere water that was so sweet and so much of it that I never saw the like'. He put these observations and reflections together: 'And then I conjectured that the visible lines of the current, and those walls of water that rose and fell in the bights with that roar that was so loud must have been the effects of the clash of the fresh water with the salt water.'[25]

This discovery complemented the growing evidence that he was skirting a huge mainland. He had not altogether forgotten the prospect of finding a southern continent. On the outward stage of the voyage he indulged in a reflection reminiscent of Pierre d'Ailly's suggestion that antipodal men might dwell in the remotest part of the Eurasian landmass, or, as Columbus put it, speaking of his own discoveries, 'These lands are another world which the Romans and Alexander and the Greeks laboured with great efforts to take'[26]—that is, lands Asiatic and 'other' at the same time. It was probably this cloud-born prodigy that was in his mind when he confided to his journal off the Venezuelan coast, 'Your Highnesses will gain these lands, which are another world'.[27] The evidence that was edging him towards recognizing their continental dimensions also made him rethink their relationship to Asia. Without necessarily revising his assessment of the size of the globe, or his conviction that he was close to the limits of the Orient, Columbus knew there was no scope in traditional cosmography for an extension of the Eurasian landmass in his present location. A new continent, however near Asia, must be distinct from it.

These considerations were impelling him towards a true understanding of the nature of America when the conviction dawned that the huge
body of fresh water he had observed must have a large extent of land to
flow through. Columbus was cautious in drawing the inescapable
conclusion, but he did ponder the assertions of a great continent to the
south, which—he now claimed to remember—Indians of the Lesser
Antilles had made on his second voyage. At last, on 13 August, when he
was off Margarita, he entered in his journal one of the most momentous
statements in the history of exploration, 'I believe this is a very large
continent which until now has remained unknown'.[28] This was not an
evanescent insight: Columbus stuck to it when he was back on Hispaniola, reporting to the monarchs his discovery of 'an enormous land,
to be found in the south, of which until the present time nothing has
been known'.[29]

Within a few days of finding the mainland of America, Columbus
had therefore correctly assessed its nature. Though he clearly
underestimated—and underestimated grossly—its distance from Asia,
he unequivocally expressed the distinction between Asia and the land he
had found. He had not of course proved that the land was 'new': such
proof was still not available until the discovery of the Bering Strait more
than two hundred years later. But Columbus had rested his credentials
as the discoverer of America on a firm basis: not only had he been the
first to find that continent in the course of a conscious labour of
exploration, and to record the find, but he had also appreciated and
articulated his achievement. It is an error to think of Columbus's
discovery as fortuitous, or to give the credit for first understanding its
nature to any later explorer. Columbus had in a sense been beaten to the
correct conclusion about his discoveries by scholars like Peter Martyr,
who had classified them as 'antipodean' from the first; but theirs had
been theoretical conjectures, based on a realistic calculation of the size
of the globe. By finding the mainland and producing evidence that the
New World included an unknown continent, Columbus had turned the
speculations into empirically verifiable facts. Columbus's scientific
breakthrough—it is not excessive to call it that—was known in his time
and quoted by early biographers. Although over the next ten years
Vespucci, Waldseemüller, and others adopted it or arrived at it to some
extent independently, and helped to make it generally known and
accepted, Columbus was in an unequivocal sense its originator.

He wanted to stay on the coast of the Tierra de Gracia and the land
which the natives taught him to call 'Paria'. The painful eye trouble he

had contracted in Cuba four years before had returned to plague him, however, and he was, perhaps, aware of his neglected responsibilities, awaiting him on Hispaniola. Confiding in the possibility of returning to explore later, he turned north and away from the coast on 15 August.[30]

Wracked by the pain in his eyes, struggling to write up his dispatches for the monarchs, he returned on the voyage north to his introspective and embittered mood. He remained as self-congratulatory as ever; his superlatives flowed as freely. Yet at the same time he included recollections of Portuguese achievements in Africa in a transparent attempt to taunt his own patrons into a livelier sense of duty. Memories of his treatment at court spilled out, perhaps, without contrivance, because they were so deeply felt. He turned to self-pity and affected contrition: it is hard to resist the impression that his reflections on the superiority of spiritual goals were an act of penance for his own material cupidity. 'I do not endure the hardships', he exclaimed, with an air of protesting too much, 'to gather treasure nor find riches for myself, for, to be sure, I know that all is vanity that is accomplished in this world, save what is to the honour and service of God, which is not to build up riches or causes of pride or many of the other things we use in this world to which we are better affected than to the things that can save our souls.'[31] Throughout the texts written on board ship or shortly after landing, Columbus, while clinging to his former hopes, seemed to ease himself away from reliance on the monarchs' patronage towards dependence on God—the blessed condition he had seen exemplified in the Franciscans he admired so much, and in the Indians of Hispaniola, when he had first beheld them with comparatively innocent eyes.

In this vulnerable mood he contemplated, without reason or criticism, the observations he had made on the voyage so far. He was in no condition to process information rationally. He recalled, first, the change in clime which he had claimed to observe on his first voyage about a hundred leagues west of the Azores. He recalled the sweet water and temperate air of the Gulf of Paria, which seemed, in retrospective, unnaturally perfect. He recalled that its river-mouths, like those of Paradise, were four in number. Lastly he adduced his astronomical observations—which were rarely accurate enough to be a source of useful insights. He seems to have worked hard at improving his stargazing talents since his first two voyages, when his purported readings of latitude had commonly been wrong by 100 per cent and his attempt to read longitude from an eclipse had produced an even wilder error. On the third crossing he had taken numerous readings of the

position of the Pole Star, and had confirmed his discovery that it tended
to deviate from the fixed position traditionally assigned to it. Now he
actually believed that he could measure the revolution Polaris described
in the heavens, though in this he was wildly optimistic. He set himself
the task of making readings of an accuracy unattainable with a simple
quadrant or astrolabe aboard a moving ship. His finding—that the angle
of elevation diminished progressively, irrespective of latitude—must
have been a delusion. Instead of blaming imperfect observations or
faulty instruments for the observed variations, Columbus, like the
doctrinaire empiricist he was, accepted the data and turned his mind to
contriving an explanation. He concluded that he must be sailing uphill:

Now I observed the very great variation which I have described and because of it
began to ponder this matter of the shape of the world. And I concluded that it
was not round in the way they say, but is of the same shape as a pear, which may
be very round all over, but not in the part where the stalk is, which sticks up. Or it
is as if someone had a very round ball, and at one point on its surface it was as if it
had a woman's nipple put there; and this teat-like part would be the most
prominent and nearest to the sky.

The perceived change of clime west of the Azores was to be explained as
the effect of the ships' beginning 'gradually to ascend towards the
heavens'.[32]

This conclusion was not perhaps as singular, or as innocently empirical, as might first appear. Columbus was also able to call on an
extraordinarily warped reading of cosmographical traditions about the
sterile problem of the orientation of the world in space, and whether the
North or South Pole should be regarded as 'at the top'. The supposed
reference to Aristotle was lifted, once again, from d'Ailly:

I affirm that the world is not spherical, but has this modification which I have
explained and which is to be found in this hemisphere where the Indians are and
in the Ocean Sea. And the highest part of it is on the Equator. And it very much
conduces to the same conclusion that the sun, when our Lord fashioned it, was
positioned above the easternmost point of the world, at the very place where the
highest point of the bulge of the world is located. And although it was Aristotle's
opinion that the South Pole, or, rather, the land at the South Pole, is the loftiest
part of the globe and the closest to heaven, there are other authorities who
disagree, stating that this is the North Pole. It therefore appears that there must
be some part of the world which is loftier and closer to heaven than the rest. But
they did not hit upon the fact that it was on the Equator in the form I have
described. And their failure should not be wondered at, because they had no
certain knowledge of the existence of this hemisphere, except for shadowy

speculations inferred by reason alone; because no one had been here or sent a mission to seek it until now, when your Highnesses ordered sea and land alike to be explored and discovered.[33]

Pierre d'Ailly had speculated, in a passage Columbus must have known, about the possibility of a protuberance marring the sphericity of the earth.[34] But the discoverer did not stop there. His speculations proceeded. The location and amenability of the delta he had discovered, with its four rivers, at 'the end of the Orient', tempted Columbus to a yet more reckless inference:

I believe that if I were to sail beyond the Equator, I should find increasingly greater temperance in the climate and variation in the stars—though I do not suppose that it is possible to navigate there, where the world reaches its highest point, nor for any man to approach, for I believe that there the earthly Paradise is located, where no man may go, save by the grace of God.[35]

Columbus had now gone quite beyond the evidence. To the world's putative breast he had added an imaginary nipple and placed the Garden of Eden on its top. With the fact that tradition located Eden in the extreme east, he had muddled two errors of his own: the supposition that he was indeed on the fringes of the Orient, and the inference from his false astronomical bearings that he was approaching the 'stalk' of the world. Future explorers of America would search for objectives no less chimerical and sometimes less creditable than the earthly Paradise, including Atlantis, the land of the Amazons, and the Fountain of Eternal Youth. But neither Columbus's new theory of the shape of the planet nor his contribution to the location of Paradise had any influence on his contemporaries.

On the other hand, it has to be said that Columbus was fertile even in error. For the world is not a perfect sphere, and it does indeed bulge towards the Equator, albeit not quite in the way he claimed. Just as his observations of magnetic variation, however inexpertly made and uncertainly interpreted, disclosed a great scientific truth, just as his contemptible small-world theory inspired a great discovery, so his quest for the earthly Paradise, by questioning the prevailing view that the world was perfectly spherical, broached the possibility of a new and more realistic perception of the shape of the planet. Nor was it necessarily impertinent to suppose that Eden might be located by empirical enquiry: in one sense, it was a great scientific *aperçu*—a bold intrusion of science into the field of faith. The location of Paradise had been an object of respectable speculation in the past and Columbus could regard it as a proper subject for the sort of contribution he was qualified to make. There was something heroic in the conviction with which Columbus believed he had trounced, by a mixture of divine guidance and scientific observation, 'Ptolemy and the other sages who wrote of this world'.[36]

On balance, of the two new hypotheses Columbus formulated on this voyage—that of the continental nature of America and that of the location of Eden—the former must be acknowledged as the more promising. The tragedy was that it became obscured by the latter. Because the earthly Paradise was to be expected in the east, the theory became inter-dependent with Columbus's increasingly desperate case in favour of the proposition that his lands were Asiatic. In an undated letter, probably of late 1499 or early 1500, he withdrew the most fruitful of his claims—the claim to have discovered an unknown continent:

The land which God has newly given your Highnesses on this voyage must be reckoned continental in extent, wherein your Highnesses must take great joy and render Him infinite thanks and abhor them who say that you should not spend money on this enterprise, for they are not friends of the honour of your high estate—to say nothing of all the souls for whose salvation we can hope, whereof your Highnesses are cause, and which is our chief gain. And I wish to address the vainglory of this world, which ought to be set at nought, because our mighty God abhors it. And let them answer me, those who have read the histories of the Greeks and Romans, if with any such little gain they extended their empires as greatly as your Highness [sic] has now done with that of Hispaniola with the Indies, and that one island which measures more than seven hundred leagues, and Jamaica, with another seven hundred islands, and so great a part of the mainland, which was very well known to the ancients, and not unknown, as the envious or ignorant like to pretend.[37]

That it was an 'unknown' land was precisely Columbus's own insight and his abandonment of it is a sad spectacle by comparison with that of the triumphant discoverer of August 1498.

'The Devil Has Been at Work'

THE HISPANIOLA COLONY,
1496–1499

THE full measure of Columbus's failure as a colonizer was not yet apparent when he returned to Castile in 1496. Yet by the end of six or seven years of his governorship, with his own, the monarchs', and the settlers' objectives all still unachieved, and Hispaniola suffering an apparently interminable series of rebellions not only by the Indians but by the colonists too, Columbus was to be superseded and disgraced, and shipped home to Spain in chains. He ascribed his ill success to the intervention of the Devil, but it is possible to discern the influence of human errors, most of them Columbus's own, as well as of intractable circumstances.

Perhaps the chief cause of disharmony on Hispaniola under his rule was the conflict of aims between Columbus and his men. He felt a personal involvement with his discoveries. He wanted to employ there men who shared his love of the place. He was resentful of anyone who did not have his faith in the merits of the land or who was unwilling to make as much effort as he on its behalf. Shortly after landing on Hispaniola in 1498, when he found the administration in disarray and a large part of the colony in rebellion, he realized that most of the Spaniards who accompanied him to the New World did not share his vision of a settler's ideal life, but were interested only in the attractions of 'the best land in the world for idlers'. He wrote to Ferdinand and Isabella:

Our people here are such that there is neither good man nor bad who hasn't two or three Indians to serve him and dogs to hunt for him and, though it perhaps were better not to mention it, women so pretty that one must wonder at it. With the last of these practices I am extremely discontented, for it seems to me a disservice to God, but I can do nothing about it, nor the habit of eating meat on Saturday [*sic*, for Friday] and other wicked practices that are not for good Christians. For these reasons it would be a great advantage to have some devout friars here, rather to reform the faith in us Christians than to give it to the Indians. And I shall never be able to administer just punishments, unless fifty or sixty men are sent here from Castile with each fleet, and I send there the same number from among the lazy and the insubordinate, as I do with this present fleet—such would be the greatest and best punishment and least burdensome to the conscience that I can think of.[1]

No more conspicuous proof of the gap in perception of the island between Columbus and his men could be furnished. There was a curious irony in Columbus's proposal to deport home for delinquency men, some of whom were already exiles for their crimes. For most of them it would have been an incitement rather than a punishment. Once the arriving colonists found how far removed were the realities of life on Hispaniola from the promises Columbus had made, it became their greatest wish to leave—even before, it may be, accumulating the wealth of which they had dreamed. 'May God take me to Castile!' was the island's favourite oath, and free passage home the first of the rebels' demands.[2]

The fault, in a sense, was Columbus's own. The picture he had painted of large quantities of gold for the picking and willing Indians to serve, all in a salubrious climate and fertile soil, had been taken literally by his public and had attracted the idlers and fly-by-nights it deserved. Conversely, it inevitably provoked disillusionment when the men found how hostile the environment really was and how great the labour demanded of them. Columbus virtually admitted all this in an epistolary threnody he sent home in May 1499:

None of the settlers came save in the belief that the gold and spices could be gathered in by the shovelful, and they did not reflect that, though there was gold, it would be buried in mines, and the spices would be on the treetops and that the gold would have to be mined and the spices harvested and cured—all of which I made public when I was in Seville, because those who wished to come were so numerous. And I knew what they were after and so had this explained to them, with all the work that men who go to settle far-away lands for the first time, and they all replied that it was to do such work that they were going.[3]

To judge from the tone of all his surviving writings about Hispaniola, Columbus's admonitions had been grossly inadequate and drowned by the torrent of his praise for an alluring Arcady. It became a not unusual sight for disappointed colonists, returning from Hispaniola, to riot before the monarchs at their public audiences, denouncing Columbus and his 'lands of vanity and delusion', which they derided with surprising acumen.[4] The large contingent Columbus had shipped out on his third crossing—with which he had hoped, perhaps, to strengthen his own hand—had only made matters worse by increasing the pressure on the colony's inadequate resources.

If Columbus's plans for the colony were out of step with those of his men, they did not fully coincide with the objectives of the monarchs, either. It is difficult at this remove to judge Columbus's intentions, but he seems, while keeping up his clamour for a stable community, to have envisaged one which was less than permanent. The model foremost in his mind seems not to have been the agrarian colony, dedicated to the direct tilling of the soil, populated by colonists at all social levels, such as existed in the Canaries, Madeira, and the Azores, but a trading establishment of the Genoese type, or of the kind the Portuguese had established at São Jorge da Mina on the West African coast, dealing in long-range exchange of high-value products, with a big surplus to be invested in the metropolis. The settlers would supervise the introduction and cultivation of European crops and the breeding of cattle, in order to furnish themselves with an acceptable diet, but would chiefly direct their efforts towards the production of gold for shipment home, with the cotton, dyestuffs, and any available spices that could be found or introduced, and of course with slaves. He clearly envisaged that the labour would be supplied by the Indians, whom he seems to have grossly overestimated as to both numbers and adaptability to work. He treated them as if infinitely expendable. He made little effort to ship out labourers, selecting rather men with technical skills—soldiers, sailors, artisans, functionaries, skilled miners, and specialized agriculturists. Among the three hundred representatives of different trades whom he took on his third crossing, only fifty labourers figured, and they were accompanied by only thirty women.[5]

Ferdinand and Isabella, however, could not be satisfied with a mere trading factory. They wanted the new discoveries to be 'peopled'—that is, colonized at all levels by their own subjects, in order to bring them firmly under the political dominion of Castile. As they wrote to the municipalities of their kingdoms in commendation of Columbus's third voyage, 'We have commanded Don Cristóbal Colón to return to the island of Hispaniola and the other islands and mainland which are in the said Indies and supervise the preserving and peopling of them, because thereby our Lord God is served, His Holy Faith extended, and our own realms increased.'[6]

The monarchs particularly desired that the land of the island should be divided among the colonists, and a new agronomy introduced, just as was being affected by their command at the same time in the Canary Islands. Above all they wished to promote the production of sugar, for which there was a relatively new and still unsatisfied mass demand in Europe. They hoped, by making land-grants and fiscal exemptions which had attracted settlers in the previous few years, to lure colonists to Hispaniola. Lastly, the elements of the new agronomy were not to be cultivated to the exclusion of the pastoral sector, which the monarchs were determined to favour in their new, as in their older, realms. They did not lose sight of the main purpose of all their provision, the extension of their own power; they therefore reserved to themselves the metal deposits and logwood of Hispaniola, and were particularly insistent that in dividing the land, Columbus was to alienate no jurisdiction from the Crown, but to preserve all legal sources of power in the monarchs' hands. These aims can be observed in the instructions issued to Columbus before his departure on the third voyage:

Whatever persons wish to go to live and dwell in the said island of Hispaniola without pay can and shall go freely and shall there be frank and free and shall pay no tax whatsoever and shall have for themselves and for their own and their heirs the houses which they erect and the lands which they work and inheritances which they plant, in the lands and places which shall be assigned to them there in the said island by the persons who through you [Columbus] shall have charge.[7]

The object of Columbus's permission to divide the land was said to be the cultivation of grain, cotton, flax, vines, trees, and sugar canes and the erection of houses and mills. Now although during his years of power Columbus indeed made a large number of land-grants, the work of cultivating the crops the monarchs desired was but little advanced. The sugar, for instance, ran wild and was not redomesticated for commercial purposes until 1503. The results were diametrically opposed to those Ferdinand and Isabella expected, particularly with respect to the Indians.

For the monarchs and the colonists were but minor elements in the elaboration of Columbus's work, as he conceived it, compared with the problems posed by the Indians. The monarchs' precepts for the treatment of this group were expressed in the first clause of their instructions to Columbus:

First, when you are in the said Indies, God willing, you will try with all diligence to inspire and draw the natives of the said Indies to ways entirely of peace and tranquillity and impress on them that they have to serve and be beneath our lordship and benign subjection, and above all that they be converted to our holy Catholic Faith, and that to them and to those who go to live in the said Indies be administered the Holy Sacraments by the clerks and friars who are or shall be there.[8]

This expression of policy was important, not only because it restated the monarchs' avowed desire to procure the conversion of the Indians, and affirmed that they would tolerate no intermediate lordship imposed by way of slave-trading or usurped jurisdiction between themselves and their newly acquired subjects, but also because it appeared to settle a controversy which would in fact embitter the history of the New World for some time to come—that of whether the Indians were sufficiently rational beings to benefit from the sacraments of the Church. Columbus's requests for friars to be sent to Hispaniola for the needs of the colonists rather than the natives were consciously ironic: he was using the simple pagan in his traditional role as a commonplace of sententious literature, to point up the moral deficiencies of the Christians. He was, beyond question, every bit as enthusiastic about converting the natives as his royal sponsors.

There was a political motive in the work of conversion. The Bull of 1496, in which Pope Alexander VI established the juridical basis of the Spanish presence in the Indies, caused Castilian rights to rest on the charge of evangelizing the natives, which he laid at the monarchs' door. In practice, this condition could have been ignored as it was by the Portuguese, who operated in Africa under a similar sanction of which they rarely took much notice; but Columbus, as well as his sovereigns, had already embraced the task with cordiality, before the Pope pronounced, as soon as the discovery was made. At the end of the 1470s, in a juridical wrangle with the papal nuncio and in the face of some unfavourable canonists' opinions, Ferdinand and Isabella had adopted the position that conversion and conquest were inseparable processes where primitive and pagan peoples were concerned. The resolution of the dispute, which arose with reference to the Canary Islanders, broadly in the monarchs' favour, brought their political interests and their evangelical aims into perfect harmony. It had not given them, however, a free hand to dispose of the natives of the New World as they pleased. On the contrary, they were bound by a long canonical tradition, as well as by their own convictions, to permit nothing such as maltreatment or indiscriminate enslavement to interfere with the work of conversion. Courses which might be commercially advantageous, such as enslaving the natives or granting them out to Spaniards as labour forces in feudal subjection, were debarred by the political necessity of preserving the Indians under the direct lordship of the Crown. In particular, it was quite clear that no natives could in law be enslaved, unless captured in the course of legitimate warfare or clearly placed outside the protection of natural law by such offences against it as cannibalism.[9]

It was at this point that the consensus of Columbus's policies with those of the monarchs ceased. The extent to which he understood their priorities is shown by the terms in which he instructed Pedro Margarit, when he left him in charge of Hispaniola during the explorations of Cuba in 1494. 'The chief thing you must do', Columbus wrote, 'is watch carefully over the Indians, and allow no ill or harm to be done to them, nor anything taken from them against their will; but rather that they should be honoured and kept in safety, so that they do not rebel.'[10] What Columbus failed to grasp, however, were the limits imposed by law and royal will on the economic exploitation of native labour. As early as 1493, when he proposed the enslavement of the Indians in the version of his account prepared for publication, a presumably editorial hand amended the proposal to specify that the slaves should come 'from among the idolators'[11]—a category whose existence Columbus had denied. The intended effect of the amendment was to make the proposal juridically defensible, since idolatry was held in some quarters to be an offence against natural law. Even after he had been back to Spain twice, and received explicit briefings from the monarchs, Columbus was still incapable of grasping such distinctions. In October 1498 he informed Ferdinand and Isabella that 'as many slaves as can be sold' could be dispatched from Hispaniola 'in the name of the Holy Trinity', and if, as Columbus estimated, there was a market for about four thousand of them, they would bring in twenty million maravedis 'at a modest price'. He went on: 'And although at present they die on shipment, this will not always be the case, for the Negroes and Canary Islanders reacted in the same way at first.'[12] Somehow, Columbus's attitude to the Indians was religious without being humane. At times the merchant in him operated to the exclusion of the Christian and the visionary.

Enslavement and export, even on the scale urged by Columbus, accounted for only a small proportion of the natives of Hispaniola. Most were required to provide labour for the colony. Again, Columbus expressed his policy best in a memorandum he addressed to King Ferdinand, after he was stripped of all responsibility in the lands he had discovered: 'The Indians were and are the wealth of the island of Hispaniola because it is they who mine and make the bread and all the rest of the Christians' food, and extract gold from the mines and perform all other duties and labours of men and of beasts of burden.'[13] Once this was recognized, the problem lay in organizing the Indians to the best advantage and, if possible, consistently with the monarchs' policy of benign treatment. The imposition of a gold tribute, which Columbus probably never intended as more than a temporary expedient, was clearly an inadequate answer to this need, because of the paucity of the Indians' gold supplies and the hardship they incurred. When he resumed effective government of the colony in 1498, Columbus attempted to organize some of the natives into work-parties to extract the gold under Spanish supervision, and to work the lands which he apportioned to individual colonists, under the powers conferred on him by the monarchs for the division of the soil. Partly as a result of these measures, a new policy was imposed on the New World, in which groups of Indians became linked not to particular tasks like gold-mining or to pieces of land but to Spaniards, who exercised over them rights which were not sovereign or seigneurial or proprietary but which were nevertheless personal. This was so thoroughly the case that the governors who followed Columbus usually bestowed not lands upon the colonists, but designated groups of Indians to serve them.

In the early years of this institution, known as the encomienda, the colonist concerned enjoyed the unlimited personal services of his Indians, although in law these rights were repeatedly revoked by the Crown and replaced by delegated rights to a share in the tribute due to the sovereign. From Columbus's day until at least the 1530s, the encomienda dominated colonial society in the New World. Desire for encomiendas dominated men from within, its operations as an institution from without. Las Casas equated encomienda-lust with gold-lust. 'The gold they came to seek', he complained, 'consisted in grants of Indians.' Most of the Indians of the conquered areas seem to have been comprehended in the system, which for a time enjoyed something of an institutional monopoly in the field of Indian policy. Except for the government at Santo Domingo in Hispaniola, the native chieftains (most of whom anyway were included in encomiendas), and the municipal councils, there was no other instrument of social, economic, and political regulation in the conquered lands. All aspects of colonial life were embraced by it. The military institutions, such as they were, depended on the encomienda-holder's obligation to serve. Upon it the hopes of successful evangelization were pinned. Commerce, mining, and industry were all linked to the encomienda system—though less so as time went on—and the link with agriculture was even stronger.

It is therefore important to consider how far historians have been right to attribute the introduction of the encomienda to Columbus. When the Spaniards came to the New World, nothing which genuinely resembled the encomienda was known in the history of their society. Grants of land, lordship, jurisdiction, and tribute all had a place in previous colonizing experience, but never before had personal services been apportioned in this way. It wildly exceeds the evidence to represent the allocation of Indians among the colonists as a delegation or an alienation of sovereign rights. The enslavement of the Indians could be described in those terms, since it effected a change in the status of the natives from royal vassals to personal chattels. But there is nothing in the surviving texts of early encomienda-grants which is incompatible with royal vassaldom; indeed, there is no evidence until the coming of the Dominicans to Hispaniola that anyone thought of these grants in juridical terms at all. All the acts did was to define the Indians whose personal services the beneficiary could make use of, and confer military and evangelistic obligations upon him. The question of entitlement was not raised: the only hint of it was in the provision that the grantee should teach the Indians the faith, since it was on that principle, enshrined in the Bull of 1496, that the legitimacy of the Spanish conquest was held to rest. It is true that Columbus's age was juridically obsessed, but not nearly so much so as some later historians. One other element helped to make these grants distinctive from anything previously known: there was effectively no mention of the limits of the services involved. There were nearly always some Indians—the youngest or the oldest—who were exempted from service, and it was usually specified that the Indians were to work on 'farmlands and estates' and mines if there were any. In other words, service was to apply to all the activities for which the Spaniards required it and there were no practical limits upon it.

Where did the idea of this institution, unique in its day, originate, if it was not worked out and applied by Columbus? The records of the Admiral's administration of Hispaniola are lost—submerged with the doomed fleet of 1502. The theory that the encomienda was another work of his inventive brain rests on the assertions of historians who wrote later in the sixteenth century, above all on the testimony of the eminent antiquarian of the reign of Philip II, Antonio de Herrera, that 'the Admiral gave inheritances or farmlands whence all the encomiendas of the Indians originated'.[14] The first part of the statement may be said to give the facts, the second Herrera's conclusion, but the one does not necessarily follow from the other. Rather it seems that in the first instance Columbus apportioned land, not Indians—land which Indians were obliged to till, the amount of land being specified and the work therefore limited. In other words, he granted land-rights with a limited right of labour use implied—not at all an encomienda proper with its omission of any rights in land and of any limitation on the services of the Indians. Las Casas seems to be getting at the same point when he says that Columbus apportioned the lands of Hispaniola and allowed the Spaniards to compel the Indian chiefs to work them.[15] Grants of this type were not inappropriate to the circumstance. Instances are recorded as late as 1508 in Hispaniola. It is worth recalling that from 1497 Columbus was expressly empowered to make land-grants. There is no evidence that he ever made grants of any other sort.

On the contrary, there are strong reasons for thinking that he in no sense introduced the encomienda. In the first place, such an action would have no precedent, whereas land-grants were consistent with almost all previous colonizing experience. In the second, it is hard to see where the encomienda or anything like it can have fitted into Columbus's plans for a trading factory with a renewable population: the encomienda would have been inappropriate, except for a colonist who chose, as they did but rarely, to stay and make his life in the Indies, or at least remain for a very extended period. Yet we have it on the authority of Columbus's companion, Michele de Cuneo, that most of his men intended to return to Spain at an early opportunity.[16] Their labour needs would have been best supplied by the organization of Indians into work-bands not allotted to individual Spaniards personally, such as Columbus is known to have established for the extraction of gold.

Lastly, the attitude of Las Casas has to be taken into account. His knowledge of the Admiral was almost unrivalled in his day. He arrived to live in Hispaniola shortly after Columbus's administration there ended and could observe the institutions at first hand. He was the bitterest critic of the encomienda system, for he believed it an insuperable obstacle to the evangelization of the Indians and an offence against natural justice. Though Columbus was his hero, he never scrupled to attack him for excesses committed at the Indians' expense, particularly for his abuse of slavery and for the destructiveness of his punitive campaigns in 1494–6. If Columbus had been held responsible for the introduction of the encomienda, it is hard to see how Las Casas could have overlooked the fact, nor knowing it have refrained from comment. Yet in the *History of the Indies*, Columbus's days appear as a golden age, spoiled only by the excesses of his subordinates, before the systematic exploitation of the Indians began. Las Casas sees Columbus as licensing an existing abuse—reluctantly and as a temporary expedient—not initiating a system.

In order to reconcile Las Casas with other accounts we do not have to ascribe to succeeding governors an institutional creativity we were loath to attribute to Columbus. There remains the possibility that the sort of relationships between Indians and Spaniards enshrined in the encomienda had already grown up casually in the days of Columbus's rule, and that the first encomienda grants made by later governors merely confirmed an existing situation. The example of Paraguay, if a distant but in some ways not dissimilar instance may be adduced, suggests that this is not impossible: there the Spanish colonists acquired personal services almost by accident—or rather, the natives supplied services voluntarily when the Spaniards took their daughters as wives or concubines.[17] There is an inconclusive but suggestive passage in Las Casas which suggests that the situation in Paraguay was paralleled in Hispaniola:

The three hundred Spaniards who were here [in 1502] . . . used by seduction or force to take the head women of the villages or their daughters as paramours, or servants as they called them, and live with them in sin. Their relatives or vassals believed they had been taken as legitimate wives, and in that belief they were given to the Spaniards, who became objects of universal adoration.[18]

It may be that here—in the casual results of fraternization with the natives—rather than in the alleged operation of a supposed juridical idea through legislation of which there is no evidence, is where the origins of the encomienda lie. This conclusion is important as an example of how the institutions of the Spanish empire in America were moulded not by 'the actions of the state', as some historians have claimed, but by factors in the environment of the New World. Columbus probably did not introduce the encomienda but looked on impotently as the social modalities created by the Spanish presence in Hispaniola—and their interaction with local customs—imposed that form on the land-grants he made and the native work-bands he organized.

The encomienda proved at best unproductive and at worst destructive. Already in Columbus's day the catastrophic depopulation of Hispaniola had set in, caused by the spread of European disease among the Indians and a presumed collapse in the birth rate. Soon it would spread to other Spanish conquests. Many contemporaries, Las Casas not least among them, felt that the encomienda with its demoralizing effects and burdensome—indeed, often cruel—application was a major cause of the demographic disaster. Few advantages seem to have accrued from it and the administrators of the empire in the next century devoted much time and ingenuity to devising alternatives to it, none of which proved entirely satisfactory either. Columbus's role in its genesis was typical of his shortcomings as a colonial administrator. The problems so far exceeded his capacities that he made little attempt to impose his will on events of which he was often a passive spectator. He met his rebels, for instance, by acceding to their demands, and when he did take a positive course of action, as in promoting slave trade, it was because he admitted he could see no alternative. He pleaded to the monarchs a string of insufficiencies—of men, of materials, of personal competence.

The abuse of the Indians might have been a means of appeasing the colonists; but appeasement feeds greed and there was a party on Hispaniola whom every concession enflamed and every curtailment provoked. When Columbus arrived back in August 1498, the rebellion that had threatened during his last period on the island had erupted and developed to the point where a rival camp—almost a rival state—had grown up in the south of the island, defiant of the *Adelantado's* commands. Columbus's delay in returning was fatal; he had succumbed to his usual temptation and made a detour to explore, instead of going straight to his destination. As a result, part of the reinforcements he sent ahead fell in directly with the rebels and exacerbated the crisis.

The rebels' leader, Francisco Roldán, claimed that the rebellion began because some of the colonists felt compelled, against orders, to disperse to look for food. The massacre which they admitted perpetrating they justified on the grounds that hunger had forced them to provoke a confrontation and they had then fought their assailants in self-defence. It is apparent, however, from Roldán's own self-exculpatory letters that they were motivated more generally by the rule of Columbus's brothers, which they resented, and the conditions on the island, in which they felt betrayed.[19] Their complaints were answered—not very effectively—one by one by Columbus on his return.[20] First, they rejected the site to which Bartolomé had transferred the settlement: 'they said it is the worst place and yet it is the best.' In fact, Santo Domingo was immeasurably superior to Isabela, and has endured to this day, which may be thought to vindicate the brothers' choice; but it did impose terrible problems of acclimatization on new arrivals. When Columbus got back, he found a third of the men laid up. Secondly, there were complaints about 'hunger': Columbus pointed out that Roldán had no difficulty maintaining his crowd of 120 rebels with more than 500 Indian servants; but it was precisely because he had abandoned the stockade and was living off the land, well away from the main Spanish force, that he was able to do so. The large concentration of inert colonists at the capital was much harder to feed, except by imports from Spain or by efforts of production which neither the Spaniards nor the Indians could or would make. Columbus's assumption that the Spaniards could forgo wheaten bread in favour of the local sort, made from cassava, is revealing of the essentially cultural problems which colonization posed. A native diet was deficient in protein in comparison with Spanish custom. To satisfy hunger from native food, Spaniards had to eat enormous quantities which defied the capacity of the local economy, geared to subsistence or very small surpluses.

Finally, the rebels resented the arbitrary powers which Columbus wielded in person and, during his absence, had delegated to the *Adelantado*. Late fifteenth-century Spain was a land of harsh justice, rarely executed. The law was cumbersome and, except in the highly popular new tribunals of the Inquisition, protective of the rights of the accused; it was applied by competing authorities, and overlapping layers of authority, which checked and balanced each other. In the little world of Santo Domingo, far from the influence of any superior official, the authority of Columbus or his deputy was untrammelled. In demanding the right of appeal in all cases to the Crown, and the suspension of judgement meanwhile, the rebels were raising a problem of fundamental importance in a monarchy which had suddenly grown beyond the reach of frequent communications, and beyond the practical reach of its traditional institutions. The particular cases which had contributed to the break-up of the colony seem trivial: Columbus had confiscated some pigs for breeding, to prevent the owners slaughtering them; Bartolomé had governed 'with such rigour', according to Roldán, 'that he put the people in such fear as caused him to forfeit all their love'.

Roldán did not accuse him of any sanguinary excesses or arbitrary deprivations of liberty. He did, however, hit on an evidently sore point with his complaint that Bartolomé had tried to deprive him of his office. Underlying the grudges and grievances, one suspects, were considerations of personality. Roldán's was another characteristic story of Columbus's inability to keep a collaborator as a friend. Like all Columbus's alienated companions, he became the epitome of ingratitude—'that ingrate Roldán, a nobody whom I had in my household'—and an agent of supernatural malevolence in a world where 'the Devil has been at work'.[21]

Columbus's first reaction, when he contemplated the disorders that had overtaken the colony in his absence, was to attempt to appease Roldán. His policy was to follow the same sort of course as on his previous visit, when he had begun by conciliating the natives, and had eventually been compelled to crush them with bloody dispassion. In late October, he wrote to Roldán as 'my very dear friend', blandly assuming that the rebel was 'waiting anxiously for my return, as if your soul's health depended on it', and conceding the rebels' first demand: a free passage home.[22] Columbus later claimed that this offer was a calculated piece of justifiable dissembling, intended to lure Roldán into custody. It may equally well be, however, that he was genuinely anxious to avoid a trial of strength with the rebels, who had ransacked the colony's arms dump, and been strengthened by 'the better part' of the new arrivals. Columbus preferred to see them in Castile, adding to the growing chorus of his calumniators, rather than in Hispaniola, directly destroying his work. In any event, the strategy failed. Roldán ignored the sailings, temptingly delayed, with which Columbus trawled for him and continued instead 'to plague me with his raids'. Not even his total capitulation to Roldán's demands and his restoration of the rebel to a post of uneasy confidence could eradicate the causes of rebellion, and after Roldán's return in honour to the Admiral's camp in August 1499, parties of rebels remained in arms.

The crisis suddenly became acute in September 1499, with the arrival off Xaragua of Columbus's old comrade-in-arms, Alonso de Hojeda. With other of the Admiral's former friends, including Juan de la Cosa who had crossed the Atlantic at least once in Columbus's company, and Amerigo Vespucci, who, as a business collaborator of Gianotto Berardi, had probably been privy to some of Columbus's plans, he mounted the first of many expeditions from Andalusia that were to infringe the Admiral's precious monopoly of transatlantic navigation. The bad news from Hispaniola, and the atmosphere of prejudice against Columbus created by the growing crowds of calumniators that thronged the court, enabled him to obtain a licence in May 1499 to look for the pearls Columbus had reported, during his third voyage, off the northern coast of the South American mainland. He traced Columbus's route, 'killing, robbing and fighting' as he went, but missed the pearls and gathered scant booty for his pains. He then turned north to Hispaniola and landed near the rebel camp, apparently determined to salvage some advantage from what had been a disastrous voyage. He put himself at the rebels' head, aroused a new Indian insurrection, and alleging that Columbus had forfeited his position by exceeding his powers, threatened to depose and replace him. Columbus, putting the poacher in the gamekeeper's place, sent Roldán against him, or—rather—left Roldán to deal with him, since the price of Roldán's allegiance was a virtually free hand in the south.[23]

The mutual confrontation of the malcontents lasted until March 1500, when Roldán managed to buy the interloper off. The seeds of rebellion continued to germinate after that. When a new native insurrection in Xaragua was provoked by a Spaniard's attempt to abduct a chief's daughter into concubinage, Roldán tried to end the trouble by expelling the malefactor. Not only did the natives continue defiant—and unsuppressed for nearly four years—but Roldán's high-handedness provided the pretext for a new repudiation of authority, led this time by Adrián de Muxica, a relative of the expulsee. In June 1500 the situation was exacerbated by the arrival off the coast of another interloper: Vicente Yáñez Pinzón, brother of Martín Alonso and another companion of Columbus's first crossing. The fear that he would intervene on the rebels' behalf proved groundless; and soon afterwards Columbus captured Muxica and put him to death. The alarming series of incidents, however, had demonstrated the apparently inexhaustible insubordination of natives and Spaniards alike.

Columbus, meanwhile, had almost run out of patience and energy. By Christmas 1499 he was near despair, and reacted in a characteristic fashion. Like other worldly afflictions, the ruin of his colony and the ineradicable curse of rebellion turned his thoughts to the consolations of religion. In a very brief but important fragment conserved in the biography attributed to his son, Columbus recorded a further apparition or experience of the direct presence of his 'celestial voice'. The impression it made on him was understandably profound, and is confirmed by an almost exactly similar account, repeated in a letter of a few months later.[24] The sense of personal contact with God was more distinct on this occasion than in the first such occurrence, on the way home from the first voyage,[25] and much less so than on the last occasion, which would occur at the worst moment of his final ocean crossing.[26] The elements of the context were always the same: Columbus was isolated and alone, spiritually and even physically, friendless and self-pitying. He was facing crisis and impending death. He was in repentant mood, abjuring his worldly greed in favour of the blessings of the next world. He felt a strong sense of the malevolence of his enemies— variously courtiers, savants, Indians, mutineers, and rebels, who all seemed to form a single, diabolic continuum. In these circumstances he felt overwhelmed by the conviction that he was enduring a test of faith; and in the access of faith that flowed from that feeling, he sensed the presence and heard the voice of heaven. Columbus's priorities had been shifting from the secular to the spiritual throughout the period of the third voyage. The experience of 26 December 1499 was a decisive moment of the process:

When all had abandoned me, I was assailed by the Indians and the wicked Christians. I found myself in such a pass that in an attempt to escape death I took to the sea in a small caravel. Then the Lord came to help, saying, 'Oh, man of little faith, be not afraid, I am with thee.' And he scattered my enemies and showed me the way to fulfil my promises. Miserable sinner that I am, to have put all my trust in the vanities of this world!

As well as driving him ever higher into the remoter transports of mystical escape, Columbus's experiences in the New World were altering his self-perception. His failure as an administrator was apparent even to himself, but he took satisfaction from his own transformation into a soldierly role, defeating ill-armed Indians or beleaguered rebel bands. It was a commonplace in Spain that 'arms and letters' were diverse, if not incompatible talents, and that both were needed for the discharge of the duties of government. Don Quixote would summarize the tradition crisply in his advice to Sancho Panza: 'You, Sancho, must dress in a mixture of a graduate's gown and a captain's armour, for in the island I shall give you to govern, arms shall be required as well as letters, and letters as well as arms.'[27] Columbus's dispatches home from 1498 pleaded repeatedly for assistance—or even replacement—by 'a well-educated man, a person equipped for judicial duties'.[28] Parvenu though he was, he was now coming to see himself as an old-fashioned aristocrat, whose great virtue was his prowess, and whose nobility was uncompromised by inadequacy in a bureaucratic role. From now on he designated himself increasingly as a 'captain'—a military term —rather than as a mariner, and his discoveries as 'conquests'. His failure in administration was a proof of his nobility rather than a blot on his competence. 'I must be judged as a captain', he pleaded, 'of cavaliers and conquests and such, and not a man of letters.'[29]

His request for a 'lettered' official perhaps only anticipated an inevitable development. All during their monarchy, it was the policy of Ferdinand and Isabella to infiltrate university-trained administrators, directly dependent on themselves, into the power-structures of locally established hierarchies. Already in May 1493 they had appointed Juan de Fonseca to administer the preparation of fleets and personnel for the Indies jointly with Columbus, and during the latter's absence in Hispaniola Fonseca had assumed sole charge of the operation. As a result, in the last year of Columbus's governorship of the Indies his monopoly of navigation was unceremoniously broken and the series of so-called 'Andalusian' voyages began in Columbus's wake, feeding on his expertise, with royal permission, to extend Spanish knowledge and sovereignty beyond the lands discovered by Columbus. Not even Columbus's superhuman efforts in exploration, which distracted him from his other duties and impaired his health, could pre-empt this inevitable development: only by freely licensing the activities of explorers could the Crown confirm and extend its hold on a New World that would otherwise become attractive to foreign interlopers.

Fonseca's role was, however, in part consciously intended as a counterweight to the potential power of Columbus. In a scarcely veiled reference to the activities of the bishop (as Fonseca now was), the Admiral asked the monarchs to appoint to offices of influence only those who felt affection for his enterprise, and not those who did all they could to hamper him and favour his rivals.

When he got the professional judge he asked for, this particular condition was left conspicuously unfulfilled. 'The man preferred', complained Columbus, 'was the very opposite of what the nature of the job demanded.'[30] The monarchs decided to appoint Francisco de Bobadilla, who combined the advantages of respectable birth and learned background (both of which Columbus lacked) with judicial powers, to deal with the rebels and investigate the grievances which the colonists had accumulated against the Admiral. He interpreted his brief primarily as a matter of preventing Columbus from doing further damage.

By the time Bobadilla arrived on Hispaniola in August 1500 Columbus was already out of patience with the Andalusian voyagers, who had defiled his discoveries—as it seemed to him—by intruding on them without his permission, alienating his natives, engaging illicitly in slaving, and sometimes landing on Hispaniola to foment rebellion. He later admitted that he thought at first that Bobadilla was such a one as these.[31] He probably therefore failed to receive him with the humility and disposition to please which he claimed in his pleas to the monarchs. Nor was Bobadilla, for his part, well disposed towards Columbus. He had come fresh from hearing the allegations of Columbus's detractors in an atmosphere at court embittered by crowds of supplicants who had returned from Hispaniola with clamours for pay and grievances against the Admiral. Equally prejudicial to Columbus's interests was a xenophobic mood then sweeping the court, where foreigners, especially Genoese, were being made the scapegoats for the problems of Castile's other colonies. Throughout the late 1490s, Genoese personnel were being dismissed from office, losing suits at the bar of royal justice, and enduring repeated legislation to limit their property holdings and sequester the surplus.[32] When applying for 'lettered' help, Columbus had already anticipated this ground of objection to his rule. 'I have been blamed in my colonizing work, as in many other matters, as a poor hated foreigner',[33] he complained, but this was not the moment to apply to the Castilian court for redress on those grounds. According to Columbus's friend, Andrés de Bernáldez, the charges against him were 'that he was concealing the gold, and wished to make himself and other accomplices lords of the island, and give it away to Genoese'.[34] Columbus's attempts to lampoon such charges and imply they were equivalent to presuming that he would 'snatch the Indies from the altar of St Peter' and 'give them to the Moors' do not seem to have evoked much mirth at court.[35] From the point of view of the friars whose specific task was the pastorate of the natives, Columbus's government was no more satisfactory. Their detailed complaints have not survived, but an abstract of them suggests that they blamed Columbus for the excesses committed against the Indians: 'he took their women from them and took all their property.'[36]

Bobadilla was authorized to take the government of Hispaniola into his own hands if he found that the Admiral had a case to answer. Among his first acts on arrival, amid a new rebellion which Columbus was trying to suppress, were the promotion of the Admiral's opponents, the clapping in irons of Columbus and his brothers, and their despatch to Spain for trial of the charges made against them. 'Is there anyone anywhere', retorted Columbus, 'who would ever regard such a thing as just?'

On the way home Columbus refused to have his shackles struck off but bore them as a badge of affected humility and perverse pride, until he was able to embarrass the monarchs by shuffling into their presence, still wearing them. The journey gave him ample time for reflection. In the letters he wrote in the course of it he adopted a new self-image as a Job-like figure, a type of patience under suffering, a model of enduring faith. He reverted yet again, as at other moments of crisis, to the dream of the Jerusalem crusade. 'As I hope for heaven, I swear that everything I have gained, even from my first voyage, with our Lord's help, shall be offered to him in equal measure for the expedition to Arabia Felix, even to Mecca.'[37] As always at times of stress, his sense of his own providential mission was enhanced, and he began a project which was to occupy much of his enforced leisure back home: the search for scriptural prophecies of his discoveries. 'Of the New Heaven and New Earth of which our Lord spoke through St John in Revelations', he implausibly assured the Prince's nurse, 'He made me His Messenger and revealed those places to me.'[38]

He also had new and more practical reflections to offer. He admitted that he had exceeded the proper limits of his authority and the law in attempting to suppress, by arbitrary hangings, the rebellion of 1500 in Hispaniola. His self-defence on that point was realistic: a savage frontier—he wrote in effect—could not be regulated with the decorum appropriate to one of the monarchs' European possessions. This admission obliged him to offer a radically new appraisal of the Indians: his disillusionment with them resounds in the picture of a wild and warlike people—'savage peoples who are bellicose and live in mountain ranges and forests'.[39] This contrasts with the admirable, almost lovable images of peaceful natives, naturally good, generated by his first encounter with them. Even Columbus, it seems, was susceptible to the sort of disenchantment evinced by his men.

Columbus's career had now touched the depths but the circumstances were not perhaps as bad as they might at first appear. Temporary imprisonment was something of an occupational hazard of office in the Spain of Columbus's day, and though he would never again enjoy the halcyon days of 1493, some redress awaited him in Castile. He had failed in his role of Oriental satrap—as he imagined himself—but was forging a new and more glorious one as a hero of hagiography, a servant of God's purposes foretold in holy writ. He retained his sonorous titles of Admiral, Viceroy, and Governor and the prospects of considerable wealth—even allowing for the non-fulfilment of many of the monarchs' promises—from his share of the revenue of the Indies. His sons were being brought up at court; the legitimate Diego, at least, was set fair to make an excellent marriage; and all his family bore the title of Don in Castile.

He was unlikely to regain power in the sense of gubernatorial responsibility in Hispaniola. One of the bitterest circumstances of his disgrace was the unanimity with which the most impartial observers— the friars of the Hispaniola mission, with whom Columbus professed a special rapport—warned the monarchs never to allow the Admiral back, for the sake of the tranquillity of the colony.[40] On the other hand, though he continually sued for reparation at the monarchs' court, the mechanics of power had never attracted him as much as glory, nobility, wealth, and the excitement of discovery. As we have seen, he was eager to share the burden of rule in his colony with someone better qualified to bear it. He felt more deeply, perhaps, the interruption of his work as an explorer—the frustration of his continuing efforts to get to the fabled Orient. The intervention of the Devil in the progress of his career occurred at a moment when he had carried his explorations to a crucial point, for on the outgoing stage of his third crossing he had become the first European since the random navigations of the Vikings to see the continent of America, and first to perceive its continental nature. In the wake of this discovery, he had undergone the most intense self-examination he ever engaged in, and given fuller expression than ever before to his doubts of his own geographic theories. First the problems of the colony on Hispaniola and then the arrival of Bobadilla had prevented him from exploring further and putting both new and old speculations to the test together. Now he was removed from the Indies and returned forcibly to Castile, while ideas he had revolved for so long were plunged anew into the heated crucible of his mind. He had pondered almost to the point of incoherence and brooded almost to the point of derangement. He had taken refuge in mystical escape and paranoid fulminations against vaguely defined enemies. He had become self-justificatory where once he had been merely self-laudatory. All these symptoms were to grow more pronounced as he faced new trials, and experienced new disasters, in the years that culminated in his final transatlantic voyage.

8

'That Sea of Blood'

1500–1504 AND THE LAST
CROSSING

ON 13 March 1500 a magnificent fleet of thirteen sail dropped down the Tagus estuary towards Belém. It was almost exactly a year since the news of Vasco da Gama's reconnaissance to India had reached the court of Portugal. Now Pedro Álvares Cabral was leading a fleet of gentlemen, on which no expense had been spared, to dazzle the rajahs and merchants of the East. The Portuguese success had been well publicized on da Gama's return. Commemorative gold coinage had been struck; a great church was rising in thanksgiving at the mouth of Lisbon's river; and the monarchs of Europe had received enviable accounts of great cities and rivers discovered, spices, gems, and mines of gold. By the time Columbus arrived home in chains from Hispaniola, Cabral was in the presence of the Zamorin of Calicut. The Portuguese had won the race to the Indies. The hopes Columbus had raised receded over the far horizon, while his career ran aground on a hither shore.

Columbus devoted his unwelcome retirement to two projects: the advocacy, with a new urgency, of his old campaign for a crusade to Jerusalem; and the cultivation of his own legend. He had first called for the liberation of the 'Holy House' before his first Atlantic crossing. He had returned to the idea many times since, in moments of stress or present trouble, as if lifting his eyes to the hills. In 1497 he had prepared for Ferdinand and Isabella a now lost memorandum about it, in which he seems to have advocated an approach by way of the Ocean Sea, taking Islam in the rear. On his way home in chains in 1500 he reverted to the idea, and sustained his campaign when home, probably—given the circumstances—from dual motives of penance and policy. For the reconquest of Jerusalem was to be, in part, a personal sacrifice, and would provide, if effective, a means of recovering his forfeit prestige.

The monarchs evidently treated the proposal coolly: in a personal letter to the Queen—undated and perhaps written at any time between 1500 and 1502—which is a frank plea for some crumb of favour, Columbus begs Isabella 'not to treat lightly the matter of Jerusalem, nor to believe that I spoke about it with any ulterior motive'.[1] He was, however, asked to justify the scheme in greater detail and towards the end of 1500 produced in reply a memorandum, 'The Reason I have for Believing in the Restoration of the Holy House to the Holy Church Militant'.[2] It is an astounding document, in which no strategy is mentioned, no practical consideration broached, no means or measures urged. His sole concern—offered amid many autobiographical digressions and complaints about the way he had been treated—is to show that God's will, manifest in scriptural prophecy, in the stars, and in the emanations of the Holy Spirit, is that Jerusalem now be recovered, for the Church, by a campaign launched from Spain. Indeed, he expressly disavows practical arguments:

I leave on one side all my navigational experience from an early age and all the discussions I have had with so many people in so many lands and from so many religious traditions; and I leave on one side all the many arts and writings to which I have referred. I rely entirely on holy, sacred Scripture and certain prophetic texts by certain saintly persons, who by divine revelation have had something to say on this matter.

Beyond this methodological declaration, Columbus seems to have had three main 'reasons' to cite in favour of the expedition to Jerusalem. The first was that the idea came from the Holy Spirit. As evidence of this elusive proposition he cited the previous interventions of the Paraclete in his life: in providing him with an education in inauspicious circumstances; in inspiring his transatlantic design; in influencing Ferdinand and Isabella to take it up; and he represented himself as a suitable vessel for the outpourings of the Spirit on the grounds that he was essentially an unlearned man, like the evangelists themselves, and like the children and innocents whom God commonly favoured as His mouthpieces. This emphasis on the irrelevance of erudition may owe something to the influence of Franciscan tradition, which assigned a high value to holy simplicity and was mistrustful of the vanity of unnecessary learning. It also clearly matched the main theme of Columbus's long-maintained war with the courtly savants, against whom he had so many times vaunted his own practical wisdom and made a virtue of his lack of learning. In his brief to the monarchs, he seemed to hesitate between two mutually exclusive self-appraisals: he repeated his oft-asserted claim to the authority that is due to practical experience, but plumped rather for a self-characterization as an ignoramus unencumbered by worldly learning of any sort, and wholly dependent on God:

It could be that your Highnesses and all others who know me and to whom this paper shall be shown will admonish me publicly or privately with various reproaches: as a man ungifted in scholarship, as a lay seafarer, as an earthly, practical fellow, etc. I reply in St Matthew's words, 'Oh, Lord, Who wouldst hide these things from the clever and learned and reveal them to little innocents!'

His second argument, compressed into a brief allusion, was that success would attend an expedition to Jerusalem if the monarchs (and their subjects, who were going to have to provide the resources) had sufficient faith: 'for in that enterprise, if there is faith, you shall have a most certain victory . . . Nothing shall be wanting for this that is in your people's power to give.' The third argument was that success was prophesied by two infallible divinatory traditions: one derived from Scripture and one derived from astrology, 'by signs in the heavens'. Columbus's conviction of the compatibility of Scripture and astrology, of sacred and pagan science, was heavily indebted, like his cosmography, to Pierre d'Ailly, whose selection of astrological authorities he followed closely. He did not get round, in his memoir for the monarchs, to citing any Scripture that foretold the recovery of Jerusalem: the texts from Isaiah to which he alluded seem to have been read by him as prophecies of the discovery of the Indies. It is clear enough, however, that in the prophecies with which he was chiefly concerned at the time of writing, the event predicted was the end of the world.

Nor was this irrelevant to his purpose. Columbus regarded the recapture of Jerusalem as a necessary precondition of that devoutly desired consummation. The establishment of Jerusalem as the court of the Last World Emperor and the cosmic struggle of that hero against Antichrist were seen, in the prophetic tradition that went back to Joachim of Fiore, to precede the last days.[3] It was therefore entirely germane to Columbus's purpose to claim, as he did, that St Augustine's calculations of the age of the world suggested that only 155 years were left to the end; and that the discovery of the New World was part of the divine acceleration of history that must precede the end; that a great mutation, including some of the expected prerequisites of the end of the world, could be anticipated from astrological data; and that Joachim himself—for he is cited by name—associated the recapture of Jerusalem with a Spanish initiative: 'he who would rebuild the House upon Mount Zion would be sure to come from Spain.'

It has to be said that Columbus was an inattentive reader of prophetic as of geographical literature. For example, a cycle of ten revolutions of Saturn described by Pierre d'Ailly had come to an end in 1489: it might therefore—with only a small stretch of the imagination—be alleged as a portent of the discovery of America, but cannot also have been relevant to Columbus's projected conquest of Jerusalem.[4] He failed to make use of d'Ailly's texts on planetary conjunctions, which could have been made to support his arguments. His references to d'Ailly's passages on Antichrist, which were highly germane to Columbus's views, were left vague and unexploited. Nor did Columbus seem aware that in those passages the Cardinal was reporting Roger Bacon. The text he attributed to Joachim seems to have been an invention of his own, or perhaps a mistaken tradition of the time.[5]

The Jerusalem project was closely allied to the second main concern of this period of Columbus's life: his effort to promote a particular image of himself. Advertising does not have to be insincere and it is entirely convincing to see Columbus as profoundly committed to the sense of divine purpose which he proclaimed to the world. He saw himself, like that other hero of his, John the Baptist, as 'a man sent from God'. It was vanity, no doubt, or at best false modesty; but a form of vanity intelligible in one who had achieved so much, yet was so little content. It was fed through some of the deepest and most rough-cut channels by which Columbus's soul was scored: anger, resentment, pride, and an ambition too easily frustrated. A weaver's boy still wistful after becoming a viceroy, an explorer still restless after discovering America: these can evoke sympathy, but only from sensitive onlookers; they are hard even for the most sympathetic fully to understand. Columbus had been encouraged, too, by the heroic role in which admirers saw him. It could have been intoxicating to be hailed as a new apostle by Jaume Ferrer or as a candidate for apotheosis by Peter Martyr.[6] All these effects—some from within him, some from without—drove Columbus steadily towards a conviction of his providential role.

If, however, there was one single, decisive influence, it has to be sought in Columbus's tendency to use religion as a refuge. He darted into the haven of religious experience whenever the storms of life threatened. His relationship with God became a substitute for his unsatisfactory human relationships. His path to heaven was picked out among the debris of his many shattered friendships. Glib about the love of God, he barely ever mentioned his wife or mistress. His trust in God contrasted with the suspicion and fear inspired by his ubiquitous human enemies. Heavenly companionship was his recourse as a deliberate loner. At times of crisis he had a 'celestial voice' to talk to, rather than a flesh-and-blood friend.

In late 1500 and for much of 1501 Columbus's self-vindicatory efforts were focused on the collection of scriptural and classical texts, supported by astrological observations, allegedly prophetic of his own achievements. He was obviously hard at work on the collection when he wrote his memorandum on the Jerusalem project: hence the curious irrelevance of the Isaiah texts, which had nothing to do with Jerusalem but which were in the fore of Columbus's mind because he saw them as self-referential. He got some help from the Carthusian Gaspar de Gorricio, who was his most prolific correspondent at the time and whom Columbus used as a fount of bibliographical advice and as a custodian of his writings. He was, perhaps, temporarily alienated from some of his Franciscan friends by the adverse criticisms of the missionaries of Hispaniola, and Gorricio acquired, for the rest of Columbus's life, the sort of combined role as spiritual father and business agent discharged at an earlier period, and still, it seems, intermittently continued, by the Franciscan Juan Pérez. The surviving texts of Columbus's collection of prophecies show the influence of no hand other than Columbus's own. A more polished version, edited by Gorricio, may have existed, or at least been intended, but all we have now is an ill-kempt mass of almost random jottings, compiled over a long time and including material added up to 1504, at least. Under the somewhat misleading title *Book or Manual of Authorities, Sayings, Pronouncements and Prophecies Concerning the Matter of the Recovery of the Holy City and of Zion, the Mount of God, and Concerning the Discovery and Conversion of the Islands of India and of All their Peoples and Nations for our Spanish Monarchs* are grouped well over a hundred entries spread over eighty-four folios.[7] Most are biblical and patristic texts about Ophir, Tarshish, and other names of special resonance for Columbus, or simply about islands. There is a fragment of what appears to be a copy of a passage on eclipses observed by Columbus with calculations of longitude arising from them, expressed in muddled and mutually contradictory figures. There is a gloss on a passage of Seneca's *Medea* (taken from a folio which survives among

Columbus's books), which Columbus quoted subsequently, suggesting a self-comparison in his mind with Jason's pilot, Typhis, 'who will discover a new world, and then will Thule no longer be the most remote of lands'. There is a tantalizing fragment of a letter purportedly addressed by a Genoese ambassador to the Catholic monarchs, attributing to Joachim of Fiore the prediction that 'from Spain will come the man who will restore the Ark of Zion'. Finally, apart from doodles and casual scribblings, at intervals throughout the collection, three fragments of verse appear. A stanza of eight lines celebrates the feast of John the Baptist, to whom Columbus, as we have seen, can be presumed to have had some special devotion. There are eight lines on the propriety of keeping one's promises, which may have been relevant, in Columbus's mind, to the suits he addressed to the monarchs. The last poem, longer and penitential in character, was drafted in the margins of the last leaf, and can therefore probably be assigned with confidence to a later period. Its introspective tone and accent on the imminence of death confirm this, but its allusions to 'accounts with Caesar', to 'malicious enemies', and to the 'pain and travail' endured by Christ's true servants, are all representative of the mood of revulsion against the world in which Columbus compiled his prophecies.[8]

The crusading project and the compilation of prophetic utterances were both conceived, in part, to influence Columbus's relations with the monarchs at a time when he was regarded as 'in disgrace, low in the favour of these monarchs, and with little money'. Columbus's own appraisal of his fortunes confirmed this opinion. He had been 'stripped of my honour and estate without cause'. 'I am not now much sought-after', he wrote to Gorricio in May 1501, 'nor does my enterprise greatly prosper.'[9] Stray references to confinement to his bed suggest that his health was also in decline. As during his past period of enforced inactivity in Spain, he plied the monarchs with advice not only on the problems of Hispaniola and of navigation to the Indies but also on every subject on which he might pretend to expertise, including the route of the Infanta Juana's voyage to join her Burgundian husband, the carriage of trade to Flanders, and the cabotage of the western Mediterranean. The most practical outlet for his talents was his constant struggle to secure his rights and revenues against the depredations of the bureaucracy and the threat of confiscation by the Crown. In this campaign, Gorricio was probably his intermediary, keeping copies of the memoranda in which the Admiral minutely listed his rights and particularized his grievances. He complained chiefly that by virtue of his dismissal he had lost profits of justice which it was his privilege to dispense; that his share of the monarchs' revenues from Hispaniola was being calculated after the deduction of royal payments to third parties; that he was being denied the right to appointment and dismissal to offices in Hispaniola which was among the perquisites of his offices, that his monopoly of navigation had been infringed, that his expenses had not been reimbursed and, in short, according to a pleading of 1501, that 'he adventured his person and those of his brothers and came from so far away to serve their Highnesses, and has expended seventeen years in it, the best of his life, without any recompense so far'.[10]

By the end of 1501 or early in 1502, there were signs that Columbus's fortunes were recovering. He was able to raise capital from Genoese bankers in Seville to take advantage of his right to invest in sailings to his discoveries—'my eighth share in the voyages of merchants who go to the Indies'.[11] Gold production in Hispaniola had accumulated to the point where a tidy sum awaited him. Above all, it was beginning to look as if the monarchs still needed him. Their policy—so detested by Columbus—of licensing interlopers to make voyages of exploration across the Atlantic had yielded, by the end of 1501, depressingly little result. The benefits of Hojeda's scheme to rob Columbus of the pearl fisheries of Margarita was reaped by the Guerra brothers of Triana, purveyors of hard tack to the Indies fleets, who were well placed to muster the capital for an expedition. The active partner in the enterprise was Pero Alonso Niño, shipmate of Columbus's first crossing, who led a pearl-fishing voyage in 1499; but they contributed virtually nothing to exploration. More adventurous but less profitable was the voyage headed by Rodrigo de Bastidas in January 1500, which reached the limit of Hojeda's navigations and explored the Gulf of Urabá before he was obliged to run for Hispaniola because of the ravages of termites. More was achieved by attempts to explore the Brazilian coast southward. Brazil was discovered by Vicente Yáñez Pinzón in January 1500 and a voyage of the following year—again financed in part by the Guerra consortium—under the gentleman-adventurer Luis Vélez de Mendoza extended knowledge of the coast probably as far south as the Rio de São Francisco. Much of the area, however, was reckoned to lie on the Portuguese side of the Tordesillas line; Portuguese navigation in the environs was laying the foundations of future exploitation. It was apparent that any new advantages for Castile would have to be sought further west and north, beyond the limits of what Columbus had already discovered. His rivals' lack of success was a powerful reason for giving Columbus a new opportunity to extend his achievements himself.

It seemed, moreover, as if Columbus could now be safely unleashed on another expedition. In September 1501 the monarchs had filled the vacancy in the government of Hispaniola, created by Columbus's disgrace, by appointing Don Nicolás de Ovando. His prestige, conferred by unimpeachably aristocratic credentials, was matched by a competence which the prosperity of Hispaniola under his rule was soon to confirm. As a former member of the Prince's court, where he kept watch over the comportment of the household, he might have been expected to know how to handle Columbus. If the Admiral were allowed to return to the Indies, Hispaniola could be forbidden him, and with Ovando in charge there was nothing to fear from Columbus's unpopularity or incompetence. For more than a year since his arrival in chains, the monarchs had endured his implied reproaches, veiled threats, and fantastic proposals. Though his writings of this period seem fascinating to us, for Ferdinand and Isabella they are more likely to have been vexatious, or simply boring. The inveterate bore is usually the last to register his own effect, but by February 1502 even Columbus was aware of it. 'I would rather be for your Highnesses a source of pleasure and delight', he protested 'than of annoyance or surfeit.'[12] By early 1502 both he and his patrons seem to have tired of it. In February, Columbus sought and obtained leave to make a further voyage across the Ocean Sea.

Before leaving Columbus let fly a volley of business letters and also wrote what amounted to a general account of his career so far, addressed to the Pope, with a request for members of religious orders to be appointed to evangelize the new discoveries. There is an unmistakable air of exhilaration about these letters, as if the tang of adventure were in the old Admiral's nostrils, reviving his hopes, rekindling his self-esteem. He told his correspondents, with ill-concealed excitement, of his forthcoming voyage. 'Another voyage in the name of the Holy Trinity', he advised the Pope, 'shall be for your glory and the honour of the holy Christian Faith'; 'I am returning to the Indies in the name of the Holy Trinity', he announced to the governors of Genoa's State Bank. The new commission had interrupted his work on the history of his enterprise 'in the style of Caesar's *Commentaries*'. He spoke, with every appearance of assurance, of the renewal of royal favour: 'My lord King and lady Queen desire to do me honour, more than ever.'[13]

Yet it is impossible to be deceived by these proofs of enthusiasm: they overlay, but do not obliterate, Columbus's continuing anxieties. The letter to the Pope, in particular, was characterized by some of the same preoccupations as were to be found in the bitter plaints of 1500 and 1501 with discovery of the whereabouts of the earthly Paradise, the projected reconquest of Jerusalem, insistence on the biblical geography of the New World, the manœuvres of Satan to frustrate Columbus's design: all these themes intruded with the Admiral's characteristic pertinacity. The discoveries were described with compressed hyperbole: the familiar expansive epithets were omitted but a barrage of fantastic statistics was called in to impress the reader: 1,400 islands, 333 leagues of the Asian coast, 800 leagues of the coast of Hispaniola. 'This island is Tarshish. It is Qittim. It is Ophir and Ophaz and Cipangu, and we have named it Hispaniola.'

It is hard to resist the impression, in Columbus's letters to Rome and Genoa, that he was preparing an appeal over the monarchs' heads to a potential rival patron, or, at least, getting a bolt-hole ready in the event of a further fall from favour. For a subject of Ferdinand and Isabella to call on the Pope to appoint clergy directly, without reference to the monarchs, was an irregular and perhaps treasonable proceeding. The Admiral broke off, moreover, with a promise to lay his complaints against the monarchs before the Holy Father privately. It is not clear whether the letter to the Pope was ever finished or sent, but the terms of Columbus's approach to Genoa raise similar speculations. By including a guarded grievance about his dismissal from his governorship, and assuring his correspondents of his undiminished filial feelings for his native city, Columbus adumbrated the grounds, at least, of a potential change of allegiance. His promise to devote part of his revenues to the relief of taxation in Genoa, if intended to curry favour with the city fathers, seems to have been effective. The reply of the directors of the State Bank included some of the most extravagant praise Columbus had received for many a day for 'having discovered by your industry, energy and prudence so great a part of the land on the other side of the world, which had remained unknown through all previous ages to people living in our hemisphere'.[14] Unhappily, Columbus did not receive this encouraging message until his return from his last and most disastrous voyage.

When Ovando left for the Indies in February 1502, his fleet numbered thirty vessels. When Columbus followed in May, he had four caravels. His brothers were with him—Bartolomé, he tells us, reluctantly impressed to serve[15]—and so was his clever son Fernando, still only thirteen years old. His commission was to explore—a more congenial purpose than the mixed bag of responsibilities he had borne on his last two crossings. His purpose was 'to go and reconnoitre the land of Paria'—to resume, that is, the explorations that had been interrupted on his third voyage. He planned, however, not to repeat the fruitless coastwise sailings of the Andalusian interlopers who had been there since 1499, but to extend the survey of the coast from Paria well to the north, verifying the northward extent of the mainland and scouting for the longed-for route to Asia. It was known by now that the continental part of the New World, which Columbus had discovered in 1498, occupied a large part of the South Atlantic. Columbus assumed that it would be possible to sail through the gap between the islands he had found on his earlier crossings and this new continent, across the western part of what is now called the Caribbean Sea. Thus he would reach the elusive land of gold and spices. At the same time, Vasco da Gama was making a second voyage to India by his proven route to the east. If Columbus's assumptions were correct, it would be possible for the two explorers to meet in the Orient, having between them traversed the whole circumference of the globe. Columbus suggested as much to the monarchs, who replied enthusiastically:

And with respect to what you say about Portugal, we have written appropriately to the King of Portugal, our son-in-law, and send you herewith the letter addressed to his captain, as requested by you, wherein we notify him of your departure towards the west, and say that we have learned of his departure eastward, and that if you meet on the way you are to treat each other as friends, and as is proper among the captains and subjects of monarchs between whom there is so much love and friendship and so many ties of blood, telling him that we have ordered you to do the same. And we shall ask our son-in-law, the King of Portugal, to write to his captain in the same terms.[16]

Columbus hoped from this voyage for the realization of his thwarted ambitions, the reversal of his failures, and the vindication of his claims, all in one crowning success. He had been sustained since his fall from favour by a mood of intermittent exaltation, by a hope clung-to and nourished by a fanciful self-perception and an unattainable project for an impractical crusade. Now he was in the grip of an old form of excitement. He was thus ill prepared mentally for the disappointments that lay in wait. For this voyage, in which he invested so much optimism, was to be the most manifest failure of his life. It would end not in Asia but in the depths of misery and near-despair. Struck by a hurricane, taxed almost beyond endurance by adverse winds, enfeebled by malaria, attacked by hostile Indians on the Panama coast, marooned at last on Jamaica, threatened with starvation, and repudiated by many of his men, Columbus was to be driven into the refuge of his declining years: wishful thinking, mysticism, and fantasy.

The crossing was the most rapid he ever enjoyed—twenty-one days from Gran Canaria. But there the plain sailing ended. He scarcely encountered a favourable wind or current for the rest of the journey. The monarchs had warned him with all the solemnity at their command not to break his journey in Hispaniola or disturb Ovando's command. They had, after all, been warned on the impeccable authority of the Franciscan missionaries that for Columbus to set foot in the colony would be fatal to the island's peace. The order was tactfully worded, 'because it is not right that this voyage should be at all delayed' and he was granted permission 'when you return, God willing, if you think it necessary, to make a short stop there on the way back'.[17]

These were not instructions Columbus was willing to obey. His sense of grievance over his derogation from the effective governorship of Hispaniola rankled badly. His legal petition disclosed, under the indifferent garb of third-person discourse, the real pain it caused him:

It is understood that it is he who has to rule and govern the said islands and lands as Admiral, Viceroy and Governor, and not another person . . . for as well as being thus stated by contract, it is just that he should do so, because of the role he has had in this business and the fact that he would not have undertaken this enterprise if it had not been understood that he would have to rule and govern, because he would not in that case have been able to hope to have any recompense from it, nor would anyone have undergone the suffering nor risked himself to bring the business to its conclusion as he has.[18]

This does not mean that he actually wanted to resume the responsibilities of his governorship: on the contrary, he much preferred exploring, but he had been afraid, while he was in Spain, that 'this enterprise is heading for ruin';[19] he had a proprietary attitude to his island and he wanted to see how the colony was getting on. He also wanted to protect his substantial pecuniary stake and—it may be presumed—to satisfy himself about the revenues from gold-extraction that awaited collection or transmission to Spain on his behalf.

Santo Domingo, moreover, was the only permanent port on the far side of the Atlantic. It was a sensible precaution, as well as an irresistible temptation, to call there before continuing with exploration. When he arrived off Hispaniola on 29 June, it became urgent to seek shelter.

A hurricane was blowing up and Columbus, who knew those waters better than anybody, accurately read the signs. He sent word to Ovando, asking to be admitted to the harbour and warning him of the oncoming storm. The Governor ignored his request and scorned his warning. 'Was there ever man born', exclaimed Columbus, 'who would not die of despair—excepting Job himself—at being denied refuge, at the hazard of his life and soul and those of his son and brothers and companions, in the very land and harbours which, by God's will, sweating blood, I won for Spain?'[20] Part of the fleet that had brought Ovando from Spain set off, unknowing, in the face of the hurricane, while Columbus's vessels sought shelter in a small natural harbour he knew nearby.

That night, while his flagship held fast, his other caravels broke their cables and survived only by luck and daring seamanship. The fate of the outgoing fleet bore out Columbus's warning. Nineteen ships were lost with more than five hundred hands. Among the dead were Francisco de Bobadilla, the official who had chained Columbus, and his old foe Roldán.[21] Also lost was the largest shipment of gold yet dispatched to Spain. The only ship to reach Castile was one that bore part of Columbus's own revenues.

Opposed by southerly winds, Columbus could not resume his explorations of the coast of Paria directly. He had no option but to head west, until, towards the end of July 1502, he came against the coast of Belize, near Bonacca at the eastern end of the Bay Islands. Justly famed for crossing the Atlantic, Columbus had now achieved a second transnavigation—that of the Caribbean—which was less spectacular but, in its way, no less remarkable. If the delays caused by the storm and adverse winds are counted, it had taken nearly twice as long to make the crossing as it had to cross the entire ocean. The complex currents, variable winds, dangerous reefs and shoals, and tempestuous weather make the Caribbean hazardous even for those who know it well; Columbus had tackled it without preparation and apparently without any native guides on board. He was so impressed with his own achievement that he felt, looking back, that he had been guided by divine inspiration—'a kind of prophecy' accessible only to the astronomically initiated—and that none of his professional pilots would be capable of finding the way again.[22]

The coast of Honduras was continental in appearance, with an obvious hinterland of high mountains. This was confirmed by the natives, who impressed Columbus as more civilized than the islanders, better clothed, with access to long-range trade and skilful in working copper. The coast trended east to west. The question was whether this land was continuous with the continent Columbus had discovered at Paria. If so, it would have been consistent with Columbus's plans to continue along his westward course and look for India in that direction. He decided, however, to turn east and sail against the wind in the direction of his former discovery (see Map 5).

Columbus averred an obligation to the monarchs to fulfil his promise of returning to 'the folk of the brazil-wood'. It may be, however, that the explanation of his conduct lies rather in misleading information picked up from the natives. He understood that on the course he adopted lay a strait which led across an isthmus to a large ocean. He made at least a tentative identification in his mind of this strait with the Strait of Malacca, through which Marco Polo had travelled at the foot of the Golden Chersonese[23] (or Malay Peninsula, as it is now more prosaically called). If this were correct, India should lie beyond. Columbus was indeed sailing along the coast of an isthmus, and the ocean which washed it to the west was the Pacific. But the strait was non-existent, and even had it been there, the Golden Chersonese would still be three or four months' sailing beyond it. Columbus had as yet no reliable interpreter. He had captured an Indian, now christened as 'Juan Pérez', whom he was training up for the job; but for the time being, his communications with the natives were by means of signs. The biography attributed to his son, which claims a special authority for this voyage on which young Fernando shipped, says that the confusion arose because native informants indicated a 'strait' or narrow strip of land— that is, the Isthmus of Panama—which Columbus took to mean a strait of water 'in its usual sense, according to his great wish'.[24] Not for the first or last time, Columbus had been led astray by wishful thinking.

For nearly four months the Admiral and his men suffered the hardships, as they chased the chimerical strait, first of a gruelling beat against the wind and later, where the coast turned southwards at Cape Gracias a Dios, of a malarial coast in wretched weather and torrential rain. The ships were damaged and the men sick. Columbus's brother Bartolomé was among the worst afflicted. In Cariai—a region around the tenth parallel on the coast of modern Costa Rica—he found rumours of civilized and auriferous lands, apparently on the far side of the isthmus. At the end of September Columbus himself was 'close to death'. The recollection of seeing his son in the grip of the malady prompted thoughts of home and self-pitying reactions:

The sickness of the son I had with me tore at my soul, and all the more for seeing him at so tender an age—thirteen years old—in so much trouble and taking it all so well . . . Another sorrow tore my heart from my breast and that was Don Diego, my son whom I had left behind in Spain, with no one to look after him, despoiled of the inheritance of my honour and estate, even though I firmly believed that your Highnesses as just and grateful Princes, would restore all to him with advantage.[25]

After a few weeks more of unrewarding and unhealthy navigation along the coast, by late October 1502 Columbus really was in gold-bearing country, the province of Veragua, near the modern border of Costa Rica and Panama. The availability of gold called a temporary halt to the quest for a strait. Columbus, wishful as ever, revived hopes that this was to be his most profitable discovery so far—a voyage to enrich his sovereigns and silence his detractors—and that 'my hard and troublesome voyage may yet turn out to be my noblest'.[26] But the malevolent weather would not allow him to make so great a conquest so easily. Suffering from fever, like many of his men, he was driven by the gales and battered by the rain further along the inhospitable coast as far as the mouth of the Culebra River:

I struggled in there, through peril and distress in plenty, to my utter exhaustion and that of the ships and men. I stayed there for fifteen days, for such was the will of the cruel weather . . . There I changed my mind about returning to the mines and decided to do something about them while waiting for weather suitable for resuming my voyage by sea. And when I had made four leagues the tempest returned. And it left me so utterly exhausted that I was barely conscious. Now my old sea-wound began troubling me again. For nine days I wandered, lost, with no hope of coming out of it alive. No eyes ever saw a sea so high and ugly and foaming. The wind would not take us where we wanted to go or give us a chance of running for land of any sort. There I remained, on that sea of blood that boiled like a cauldron on a huge fire. So fearful a sky was never seen. One day it burned all night like the fire in a stove and cast such flames, in the form of lightning flashes, that at every moment I had to check to see if masts or sails had been struck. They flashed so fast and furious that we all believed the ships would sink on me. In all that time the waters of heaven never stopped pouring down. It was not what you would call a rainstorm, rather the coming of a second Flood. The crews were so far ground down by this time, that they longed for death as an escape from so many sufferings. The ships had now lost boats, anchors and cables twice over. They were badly holed and stripped of sails.[27]

Columbus no doubt exaggerated his tribulations, or depicted them over-vividly under the influence of fever and continuing adversity, but it was late November before he was able to begin his return to Veragua.

This perhaps suggests that he was being disingenuous in his claim that he felt obliged to return there because of the weather: only the lure of gold can have justified the feats and endurance described. Even now adverse winds and bad weather continued to impede him, so that it was not until the Feast of Epiphany, 1503, that the little fleet arrived at the mouth of the river which Columbus called Belén (Bethlehem) in honour of the day. Veragua proved to be a poor reward for so much suffering. At first, the signs were good. The Guaymis Indians appeared friendly and willing to trade. An expedition under Bartolomé Colon found rich gold deposits upriver. In fact the gold was plentiful but the steepness of the place and the volume of rain made it impossible to get at. The Spaniards soon felt the deterrent effects of torrents of water that cascade down the mountainsides, washing all material before them and even striking the sea and breaking against the Spanish ships with such force that their cables were strained or sundered.

What was worse, relations with the Indians rapidly deteriorated—as always happened when the cupidity and rapacity of the rank-and-file explorers were revealed. Columbus intended to leave his brother and a garrison in a fortified post on the banks of the River Belén, to open trade and prepare large-scale exploitation of the gold, while the Admiral and the rest of the fleet would take the news to Hispaniola and mount the shipment of tools and supplies from there. But the Indians turned threatening, while low water penned the ships inside the estuary. There was no refuge on shore and no escape by sea. After so long afloat in torrid waters the ships were holed by termites. No one knew how long they would remain seaworthy. In April, at the blackest moment of this long ordeal, ideal conditions—isolation, despair, fever—recurred for one of Columbus's experiences of the presence of God.

By April, my ships were all riddled with worm and I could barely keep them afloat. At about that time, the river opened a channel to the sea and, by a tremendous effort, I managed to float out three of the ships, empty of stores. The ships' boats went back to load the salt and water. The sea rose and turned ugly and they could not get back out. Natives appeared and attacked the boats in force but they were killed. My brother and all my other men were aboard a ship that was still inside the estuary, and I was utterly alone outside, on that wild shore, with a high fever upon me. I was utterly drained, dead even to the hope of escape. That was the state in which I dragged myself up the rigging to the height of the crow's-nest. I clamoured and cried aloud in fear, imploring, urgently imploring the captains of your Highnesses' men-of-war, from every corner of the wind, to come to my help. But never an answer returned. Still groaning, I lost consciousness. I heard a voice in pious accents saying, 'Oh, foolish man and slow

to serve your God, the God of all! What more did he accomplish for Moses or for
his servant David? From the hour of your birth He has always had a special care
of you. And when He saw that you were of an age that it pleased Him, He made
your name resound in all the earth. The Indies, that are so rich a portion of the
world, He gave you for your own. You bestowed them where you pleased, and
He empowered you to do so. Of the bonds of the Ocean Sea, which had been
bound with such mighty chains, He gave the keys to you. And you were lord over
many lands, and your honour was great among Christian men. What more did
He do for His people Israel, when He delivered them out of Egypt? Or for
David, when from a shepherd He made him a king in Judaea? Turn to Him and
confess your fault. His mercy is everlasting. Old age will not make all great deeds
impossible for you. Manifold and great is the inheritance in His gift. Abraham
surpassed a hundred years when he begat Isaac. Nor was Sarah a young woman.
You are calling on God to help you: consider, in your ignorance, who has
afflicted you so often and so sorely—God or the world? The privileges and
promises that God grants, He does not break. Nor does He say, after service has
been done Him, that such was not His meaning and that His words should be
otherwise understood. Nor does He bestow a martyr's lot as a means of cloaking
compulsion. He abides by the literal sense of His words. Whatever He promises
He bestows with increase. That is His way. As I have told you, so has your
creator dealt with you, and thus He deals with all men. And now', the voice told
me, 'show Him the resolution you have shown in all your endeavours and angers
in the service of others.' And, half-dead as I was, I heard all this, but I knew no
way to respond to those words of truth, save to weep for my sins. Whoever it was
Who spoke, closed with the words, 'Be not afraid, but of good courage. All your
afflictions are engraved in letters of marble and there is a purpose behind them
all.'[28]

This was Columbus's longest and most detailed account of his experiences of his celestial voice, either because on this occasion the voice had
most to say, or because on this occasion the circumstances were most
traumatic, and the voice most therapeutic. The nature of the voice here
took on a new characteristic as an ally in Columbus's self-projection,
and a mouthpiece of some of his bitterest complaints against Ferdinand
and Isabella. The Indies and the Ocean, for instance, were depicted as
bestowed by God directly on Columbus in person, as a domain which
Columbus, by divinely delegated authority, passed on to the monarchs.
Not only was this a distortion of the facts, since it was by grant of the
monarchs that Columbus derived his authority and not the other way
around, it was an extremely subversive doctrine: the legitimacy of
medieval government depended on the assumptions that the powers
that be are ordained of God and that divine election elevated monarchs
above their subjects. For anyone but a king to claim his honour directly

from God was, by Columbus's time—though there were earlier instances—an unimaginable challenge to royal authority. For much of its speech, the voice used language heavily saturated in biblical and liturgical allusions. Its credentials were established by stock phrases like, 'Be not afraid, but of good courage' and 'Turn to Him and confess your fault'. In particular, the voice borrowed from Job and Jeremiah— books which Columbus thought of as in some sense applicable to himself—and from Christ's eschatological discourse, which might have been assiduously perused by a reader like Columbus, who was hoping to play a part in the recapture of Jerusalem and gleefully anticipating the end of the world. The voice selected, as models for Columbus to follow, an impressive list of scriptural figures traditionally regarded as types of Christ. There was, however, a passage of two sentences in the middle of the speech where the message was stripped bare of biblical clothing. It occurred just after the mention of Abraham had given the voice a cue— as it were—for introducing the concept of a covenant. 'The privileges and promises that God grants, He does not break' seems to echo 'The Lord has sworn an oath He will not change' although the term 'privileges' updates the allusion and raises by implication the very point at issue between Columbus and his patrons. Then follow the two tell-tale sentences. 'Nor does He say, after service has been done Him, that such was not His meaning and that His words should be otherwise understood' was a direct rebuke to Ferdinand and Isabella: conveniently for Columbus, the voice spared him the need to accuse them of bad faith in so many words. The same complaint underlay all the petitions Columbus addressed to the Crown in the last years of his life: he was not receiving the rewards which the monarchs had undertaken to give him in 1492. The voice's next sentence was obscure, but its implication seems to be that God accepts only voluntary martyrdom, whereas Columbus had it thrust upon him by the cruel monarchs.

The voice seemed so exactly to articulate Columbus's own thoughts, in a way so convenient to the explorer's purpose, that a critical reader is almost bound to wonder whether the whole apparition was not a sham, a deliberately contrived device for the transmission of a lightly coded and risky 'sub-text'. Of all the coin in the critic's pocket, sincerity is the hardest to assay, but it is worth bearing in mind a few reasons for thinking that Columbus genuinely believed in his voice. This was, after all, the third occurrence of the apparition; a single instance would excite suspicion; frequent reappearances might suggest over-exploitation; but three such experiences are neither too few nor too many to be easily credible. On each occasion, Columbus's descriptions become more explicit and more detailed: one explanation for this could be growing confidence in the presence of a disturbing phenomenon which was becoming gradually familiar. Each time, the circumstances were similar, and conducive to a mystical experience in someone whose sensibilities are suitably attuned: on this last occasion, Columbus particularly mentioned that he was actually feverish, as well as isolated and despairing, when the voice spoke. Finally, apparitions of various sorts—usually physical manifestations of the Virgin, but including 'voices' and the visible descent of other celestial beings—were part of the fabric of popular religious experience in Columbus's day: not commonplace, but not freakish either, and, on the whole, still welcomed and even actively promoted by the Church. Official attitudes changed in Spain from the second decade of the sixteenth century, when visionaries became objects of scepticism and even suspicion;[29] while the beginnings of that change can perhaps be detected in Columbus's day, the explorer lived in a world where interventions from on high commanded general respect. Those usually sceptical about close encounters of this kind may be willing to acknowledge that Columbus's capacity for self-delusion was so great and the mystical experience—with its messianic discourse and outrageous egotism—so characteristic that the entire episode has a ring of truth.

Once he had recovered, Columbus succeeded in extricating his brother and the short-lived garrison before the Indians could finish them off. As on his second voyage, he had witnessed the collapse of his hopes, which had brought on a condition resembling temporary derangement, but from which he recovered sufficiently to rally himself and his men and escape from immediate danger. The problem of the worm-eaten ships' hulls, however, made escape fraught with new danger. The only hope seemed to be to make the shortest possible run to Hispaniola with the greatest possible speed. Pumping and bailing, the explorers worked their way eastward along the coast. Though his cosmography was wildly inaccurate, Columbus's navigational sense was unimpaired. He realized they still had a considerable easting to make before they could get upwind of Hispaniola. His prestige in the fleet, however, cannot have been very high and he admitted to a disagreement with his pilots. The consensus was that they were much further east than he estimated, and that they ought to turn away from the mainland as soon as possible. The men must have been afraid of being marooned on the South American coast; nor were these circumstances in which

Columbus's devious techniques of captaincy can have inspired complete confidence. He could have been suspected, for instance, of attempting to persevere in his mission to the coast of Paria, or of still trying to look for the 'strait' to India. In fact, by the time they left the coast on the Darién peninsula on 1 May 1503, they had reconnoitred almost as far as the previously known northern extension of the South American continent. Columbus had effectively proved that the mainland must indeed be continuous from Honduras to Brazil. It was the last achievement in the most impressive record of discovery any explorer had ever achieved.

His description of the nightmarish run north is, perhaps, too dramatically satisfying to be believed. He had only two leaky ships left, 'more holed than a honeycomb, with the crews in panic and despair'. Their first effort to get away from the coast was thwarted by storms and they put back under bare poles. Three anchors were lost on the flagship when 'at midnight, with the whole world around me in a state of flux', the cables snapped on the companion-vessel, which loomed up threateningly. 'By some miracle, we were not smashed to smithereens.' The second attempt was no more successful, though this time they were blown back into a safe anchorage. After eight days' further delay, they at last got away from the coast, 'still with adverse winds and the ships in worse case than ever'. They worked with pumps, buckets, and cooking-pots to stay afloat. The memory of them stirred Columbus's contempt for the armchair navigators he had to contend with at home:

Let us hear what their comments are now—those who are so ready with accusations and quick to find fault, saying from their safe berths there in Spain, 'Why didn't you do this or that when you were over there?' I'd like to see their sort on this adventure. Verily I believe, there's another journey, of quite a different order, for them to make, or all our faith is vain.[30]

As Columbus had predicted, they had turned north too soon. Their course brought them up on the coast of Cuba. A last desperate effort to reach Hispaniola before the perforated vessels should sink landed them short, for the wind was against them, on the Jamaican coast, at the modern St Ann's Bay.[31] They were castaways, in effect, since their ships were unusable and the nearest Spanish settlement was over 450 miles away, including more than a hundred miles of open water, at Santo Domingo. The distance was easily negotiable by canoe, but it is a curious feature of Spanish exploration of the Caribbean that it owed little to existing native knowledge; indigenous culture was underestimated and native craft mistrusted. Later in the sixteenth century canoes became a basic means of carrying Spanish trade, and it was to be by means of canoe that Columbus's rescue from Hispaniola would ultimately be effected. At first, however, the idea of recourse to so simple and effective a form of transport seems not to have occurred to the castaways. Their priority was survival. They were dependent for supplies on what they could get from the natives by barter and were restricted to a dyspeptic diet of rodents' flesh and cassava. Spaniards in the New World had a poor record in sustaining friendly relations with natives. The problems of excessive demand and insufficient barter soon left the castaways hungry and the natives alienated. Columbus showed all his old resourcefulness and ingenuity in the crisis. An eclipse was due on 29 February: naturally, he was awaiting it with some relish in the hope of timing it and so checking his presumed longitude.[32] By predicting it, he prefigured a hero of Rider Haggard's and intimidated the Indians into renewing supplies.

Similar abilities were called forth by the outbreak of a rebellion among a group of his own men. The Porras brothers, Francisco and Diego, had been nominated to commands with the expedition by the royal treasurer, Alonso de Morales. Columbus had shipped them reluctantly and they had never got on. They conceived a plan to return to Hispaniola by canoe with other malcontents; Columbus had little option but to let them go, but they foundered and were reduced to living by terrorizing the Indians. The Admiral had few means of coercion at his disposal, short of shedding blood. His powder and shot were exhausted and his loyal crew's morale presumably fragile. He attempted to deal with the brothers as he had dealt with Roldán—by cajolery and appeasement. They rejected his approaches. Their indiscipline endangered the whole expedition by threatening the delicate balance of Columbus's dealings with the natives. Columbus ascribed the initiation of hostilities to the mutineers:

The Porras brothers returned to Jamaica and sent to me demanding that I should hand over all my supplies, or else they would come and get them and it would cost me dear and my son and brothers and all the other men who were with me. And as I did not accept this demand, they—to their undoing—set in motion the execution of their threat. There were many dead and wounded enough; and in the end our Lord, Who hates pride and ingratitude, delivered the whole pack of them into our hands.[33]

Another arbitrary exercise of justice—such as had got Columbus into trouble in the case of Adrián de Muxica—was not to be risked.

Columbus let the mutinous brothers go. Later, the Governor of Hispaniola was unwilling to proceed against them and they were able to join
the chorus of Columbus's detractors in Spain. 'The delinquents turned
up at court, brazen-faced', Columbus fulminated in a letter to his son.
'Such effrontery, such wicked treachery was never heard of before.'³⁴

In the mean time Columbus had taken steps to procure the expedition's salvation. His loyal subordinate, Diego Méndez de Salcedo, was
one of the most daring adventurers of the expedition. He had given good
service in negotiations and action with the Indians on the mainland and
in Jamaica. By his own account, the disaffected complained that he was
Columbus's favourite, to whom the Admiral entrusted 'all the jobs that
confer honour'. His new task was to reach Hispaniola by canoe with
native oarsmen, accompanied by a second canoe with the Genoese
officer Bartolomeo Fieschi aboard. The distance was puny judged by
present-day standards of seamanship and endurance in small craft, but
Méndez spoke the thoughts of all when he replied that it was impossible.
He accepted the mission with a convincing display of reluctance, only
after a general call for volunteers had failed.

Then I rose up [he remembered] and said, 'My Lord, one life have I and no
more. I am willing to adventure it in your Lordship's service and for the welfare
of those present, because I have hope in our Lord God that, seeing the good
intention with which I shall do it, He will deliver me, as He has done many times
before.' When the Admiral heard my decision he arose and embraced me and
kissed me on the cheek, saying, 'I well knew that there was none who would dare
to undertake this enterprise but you.'³⁵

Under the surface of Méndez's self-dramatization there is a genuine
pride in what was seen as a heroic act, which he recorded in his will and
had engraved upon his tomb. The hazards of the journey almost
justified his apprehension: the canoes laboured against the current; the
fresh water gave out; the oarsmen began to die of thirst and Méndez fell
sick. At last, after five days' toiling at the paddles, both canoes reached
Hispaniola safely, though at a point far distant from Santo Domingo.

Such leisure as Columbus had on Jamaica, in the intervals between
the struggle for survival, the war against the mutineers, the attempt to
send for aid, and the dealings with the natives, was devoted to geographical reflection, and therapeutic forward-planning. Understandably, perhaps, in his desperate circumstances, under the influence of his failing
hopes, neither activity was controlled by much healthy discipline. He
was steeped in barely rational self-pity, aggravated by a genuine sense of
danger:

My plight is as I have described it. So far, I have wept my own tears. Now let the skies pity me and the earth weep for me! Of worldly goods, I have not so much as a mite for the offertory. Of spiritual comforts, here I am in the Indies, bereft as has been said. I am marooned amid this terrible sorrow; sick; waiting day by day for death; surrounded by a million savages, who seethe with cruelty and hostility towards us; and I am so cut off from the sacraments of holy Church, that my soul shall be forgotten if it leaves my body in this place. I implore the tears of all who love charity, truth and justice.[36]

As always in adversity, the elements of the old syndrome flowed from Columbus's distraught brain in a flood, mingled this time with the fear of old age. He had, he said, been to within ten days' journey of the River Ganges; he had seen the gold-bridled horses of the Massagetes, who dwell near the Amazons; he had narrowly evaded bewitchment by their sorcerers; in Veragua he had discovered the mines of Solomon; now he would go on, not only to recover Jerusalem but also to convert the emperor of China to Christianity. He repeated his claim to have discovered the earthly Paradise.[37] The most disturbing feature of Columbus's mental state at this time was that he sought to circumvent the fact of his failure in the search for a route to Asia by senselessly affirming that he had succeeded. All the mistakes he had made, and even the errors he had discarded, he now reasserted with the insistence that he had been right from the first. His own great achievement in recognizing the true nature of the American continent and exploring its northwards extension as far as Honduras was submerged beneath the fallacious assertions that Cuba was a part of China, that all his discoveries were Asiatic, and that all the elements of his original theory about an Atlantic crossing, even down to the false value assigned to the length of a degree, were right. 'This world is small', he declared, '. . . Experience has now proved it. . . . The world, I say, is not as big as the common crowd say.'[38] While good luck made him lucid, Columbus had been able to transcend the effects of his mistakes to show his genius as a navigator and his deserts as a discoverer. His writings from Jamaica with their obstinate reversion to falsehoods mark the virtual end of his intellectual development, the triumph of obsession under a malignant star.

He wrote them in no certainty that he or they would survive. As the months dragged on with no word from Diego Méndez, the prospects for the castaways must have looked bleak. Méndez had struggled through to Santo Domingo, but had found Governor Ovando predictably unsympathetic to Columbus's plight. He refused to release ships for an immediate rescue, and obliged Méndez to await the next fleet from

Spain and hire vessels on his own account. Only after seven months of officious procrastination did Ovando send Diego de Escobar to verify Columbus's plight: this was an inauspicious choice of emissary, for Escobar was a veteran mutineer who had joined Roldán's rebellion on Hispaniola in 1499. By the time he arrived, however, even the sight of an enemy was welcome to Columbus, for it showed that Méndez had got through and that help would eventually materialize.

The letters Columbus sent back to Ovando were, in the circumstances, models of tact, insisting on his gratitude and confidence in Ovando's good offices, complimenting the Governor on his elevation to a grand commandership of his order of Alcántara, and suppressing any temptation to complain at the postponement of his rescue.

If I ever spoke a true word, such is what I am about to say: that from the time I met you, my heart has always been satisfied with whatever you have done for me . . . I have neither the wit nor strength to express how convinced I am of it. Let me only say, my Lord, that my hope has been and is that you will not spare yourself to save me, and I am certain of it, for so all my senses inform me.[39]

Las Casas praised these words as evidence of Columbus's epistolary simplicity and artless character.[40] They seem, on the contrary, to be carefully contrived and highly disingenuous.

The rescue vessel arrived in June 1504. The castaways had languished for almost exactly a year. They had escaped death from shipwreck, starvation, hostile Indians, and militant mutineers. Peter Martyr likened their plight to Achemenides' among the Cyclops.[41] The official report of their exploits to the royal council back in Spain, written by Diego de Porras in his capacity as scrivener, said simply, 'We were in Jamaica doing no service. The reason for our going to Jamaica, no one knows, except for sheer caprice.'[42]

'The Messenger of a New Heaven'

DECLINE, DEATH, AND REPUTATION

WHILE Columbus languished in Jamaica, Queen Isabella's health was in visible decline. At times she could not see to write, at others was too weak to walk. She withdrew from the labours of government gradually, as the strength drained from her, and the use of her name dwindled from the superscriptions of royal documents. No one dared foretell her death but everyone knew it was impending. The tension and alarm always current at a monarch's demise began to grip Castile and the great nobles 'sharpened their teeth like wild boars in the expectation of a great mutation in the state'.[1] By the time the discoverer of America arrived home, Isabella was dying. On 12 October 1504 she made her will, her mind wandering back to the vows she had broken to gain the throne and the laws transgressed to keep it. On 26 November she died in Medina del Campo, and her unembalmed corpse began a long journey through driving rain, under blackened skies, to her mausoleum in Granada.

Columbus felt strongly that he had lost a special protectress; yet he did not at first seem apprehensive about dealing with the King alone. As soon as he heard the news of Isabella's death he wrote in suitable terms to his son Diego, who was at court, attempting to manage the campaign for his father's rehabilitation and reward. Columbus's first injunction for his son was 'to pray earnestly and devoutly for the soul of our Lady Queen'.

Her life was ever catholic and saintly and she was exacting in all that pertained to God's holy service. Therefore we can be confident that she has gone to glory and is free of all the concerns of this harsh and wearisome world. The next thing is to be vigilant and diligent, in all and for all, in the service of our Lord King and to strive to spare him from adversity. His Highness is the head of Christendom. You know the old saying: when the head sickens, so do all the members. That is why all good Christians should pray for his life and health, and those of us who are committed to his service should contribute with all diligence and zeal.[2]

Columbus still nurtured hopes of the royal widower. The choice of words in which he styles him 'head of Christendom' a style repeated in very similar terms in another letter of about the same period—might even suggest that the Jerusalem project was again in Columbus's mind; for it was one of the proper styles of the 'Last World Emperor' of prophetic tradition. Columbus talked with every appearance of confidence of being sent back to Hispaniola as governor and loaded his letters for the court with detailed recommendations for the colony.

Over the next few months, however, his mood changed; his confidence vanished; his trust in the King ebbed. He was enjoying unwonted material prosperity because of the remittances in gold which reached him from Hispaniola; but he displayed, more and more, the outward appearance of a broken man. His health was in ruins. During his last voyage he had consistently spoken of himself as an old man, with 'not a hair on my body that is not white' and 'my health broken'. By the time he got back to Spain he could only move on a litter; the journey from Seville to the court had to be repeatedly postponed and contemplated only in easy stages 'for this trouble of mine is so painful, and the cold aggravates it so, that I shall be unlikely to avoid ending up in some inn'.[3] The last remark was not entirely jocular: Columbus meant that he might die on the way. In his personal absence from court, his affairs were handled by Diego—supported from December 1504 by Bartolomé and young Fernando—and by Diego Deza, who became Archbishop of Seville in January 1505; their combined efforts yielded few or no results. The issues at stake were, as ever, the terms of Columbus's privileges: the patronage and revenues the monarchs were unwilling to concede, the offices Columbus was incompetent to discharge. Early in 1505, Columbus expressed his despair to Deza in a letter which blames the King, with shocking frankness:

And since it seems that his Highness is not willing to comply with what he has promised with his word and in writing jointly with the Queen—God rest her soul—I feel that for a simple ploughman like me to battle on with him against me would be like beating the wind. And it will be for the best. For I have done what I can; and now let it be left to the Lord our God to do it, Whom I have always found very favourable and a very present help in trouble.[4]

If Las Casas's testimony is anything to go by, Columbus seems to have conceived the notion not only that the King was hostile to him, but also that the monarch was bent on depriving Diego of his inheritance.[5] If we are right in thinking that dynastic ambition had been the motor of Columbus's life, it is understandable that anxieties of this sort would have dogged the dying Admiral, who had forgone so many triumphs. His fears, however, seem to have been groundless. Though the King was in no hurry to satisfy Columbus, that was less because of personal hostility than because delay was the Crown's standard resource in dealing with petitioners. Ferdinand was really remarkably friendly to the interests of Columbus's family. For example, he saw to it that Diego received a pension of 50,000 maravedis a year; he used royal clout in the marriage market to bring the boy a brilliant match with Doña María de Toledo, who was the Duke of Alba's niece and the granddaughter of the hereditary Admiral of Castile; and eventually, after an interval of years, he restored Diego in the office of Governor of Hispaniola, fulfilling one of the old Admiral's last hopes. Thanks to the King's matchmaking, Columbus became the posthumous progenitor of a race of dukes, and the house of the Admiral of the Ocean Sea was allied with that of the Admiral of Castile. On the matter of Columbus's pecuniary rewards, however, the King was unyielding. Las Casas depicted an interview in the early summer of 1505, when Columbus, having recovered barely sufficient health, rejoined the court by a painful muleback ride, and appeared before the King in Segovia. He acrimoniously declared that he would return his letters of privilege and retire to some sequestered spot in which he might have rest. The King replied with his usual re-assurances: he would give Columbus his due and more besides—an empty promise, since it was precisely over the matter of defining what was due that they differed. The discoverer then withdrew, somewhat consoled, explaining to himself that the King was unable to settle the affair definitively while the heirs to the throne of Castile, King Felipe and Doña Juana, were absent from the kingdom.[6] What might have passed between the interlocutors on an occasion such as this is suggested by a fragment of a letter from Columbus to Ferdinand probably of about this date.[7] It begins with Columbus's now familiar assertion that 'it was by a miracle that our Lord God sent me here to serve your Highness'. An unfavourable comparison between Ferdinand and King João II of Portugal follows, who 'took personal responsibility for matters connected with exploration, rather than delegating them to anyone else'. The implied aspersion was cast on Bishop Juan de Fonseca, whose interference had never been congenial to Columbus. The Admiral continued with tactless reminders of his supposed opportunities for service with other patrons before returning to his insistence that 'my enterprise . . . is, and will prove to be, what I have always claimed'. Most of the rest of the letter restates Columbus's claims to the specific performance of all his patrons' alleged promises, before closing:

And if I am restored to favour, you may be sure that I shall serve you for these few remaining days that our Lord shall grant to me to live; and that I hope in Him, as I feel in my heart and seem to know for certain, that I shall make that service which I still have to perform resound a hundred times more loudly than that which I have done.

Confrontation with this sadly declined Columbus—sick, dying, importunate, reproachful, and vainglorious all at once—is unlikely to have seemed agreeable to the King. Columbus's insistence on his personal election from on high may have been irksome; his recollection of his opportunities to work for other princes could have been interpreted as an implicit threat; his promise to render more glorious service in the future can hardly have been credible. These themes in the Admiral's last supplications to his patron, together with the renewed proclamation of his conviction that he had been right about his discoveries from the first, conjure up the mental world into which Columbus had withdrawn: incorrigibly self-righteous, implacably defiant, inaccessible to reason.

And yet, conscious of approaching death, he was also resigned. In the margins of the last leaf of his collection of prophecies, he wrote, probably in his last months of life, a long and self-referential poem: the fair copy he began to transcribe was never finished.[8] In its ballad-like form and sententious message the poem resembles the genre, popular at the court of Ferdinand and Isabella, of moral philosophy in verse, in which Franciscan writers specialized.[9] Its quietism is disturbed by some of Columbus's pet obsessions—the malevolence of his enemies, the balancing of his accounts with 'Caesar'.[10] Each verse begins with a Latin term; read sequentially, these yield the sentence: 'Memorare novissima tua et in aeternum non peccabis.' In the Middle Ages this saying was associated with images of St Jerome and was interpreted as an exhortation to penance: in this tradition, it could be translated, 'Be mindful of thy most recent actions and thou shalt avoid sin in eternity.' For Columbus, the phrase, in context, conveys a different, triumphant and self-righteous meaning, which I have tried to bring out in a translation of what is, after all, virtually Columbus's own epitaph. The terms originally in Latin are capitalized. They form what for the author was a self-referential sentence; and the entire poem expresses the mood in which Columbus approached his end:

> REMEMBER, Man, in time of trial,
> To hold, whoever thou may'st be,
> Steadfast to God without denial,
> If thou wouldst reign, within due while,
> With Him in immortality.
> Our end in death we all shall see.
> Give thought to making thy provision
> To clear the way for thy last mission,
> When the time comes to sail that sea.
> UNHEARD-OF DEEDS have been provided
> By holy saints, time and again,
> Who fled the world, its ways derided,
> Upon Christ's service ay decided.
> By travails wracked, enduring pain,
> They spurned the motley and the strain
> Of flesh, which is all vanity.
> So thou in due humility
> Must now thy passion's frenzy rein.
> OF THINE OWN deeds the contemplation
> Must be thy very urgent case,
> And if thy wrongs shall lead apace
> To wicked men's last destination
> Or whether to that joyful station
> Attained by just men, who have paid,
> To God and Caesar, duly weighed,
> Their ultimate consideration.
> AND thou must raise uplifted thought
> Henceforth to heaven and must flee
> The coarse world's dull depravity,
> In wisdom seeking glory's court,
> Ever resolved to set at nought
> The vicious sins that would enslave thee.
> Follow the counsel that shall save thee
> And learn to shun the other sort.
> EVER shall they joyful sleep
> Who embraced good without alloy
> And ever, too, shall others weep
> Who feed the fires of the deep.
> Because their lives they did employ

Maliciously, and did enjoy
The pleasures of the world and greed,
Forever forfeit is their mede
Of riches that can never cloy.
SINLESS BE, and contemplate
The agonies of those who die,
How grief and terror are the fate
Of sinners in their wretched state.
Think well, as far as in thee lie,
Upon the just, released at last
From travails suffered in the past,
Into the light eternally.

In August 1505, soon after his last recorded exchanges with the King, Columbus, who had often claimed to see death in the offing, felt its imminence again: only now, with a new urgency. On the 25th of that month, he wrote a hurried codicil to his will with his own hand.[11] As usual, when he sat down to write for posterity, he allowed his thoughts to wander back over his career; and, as usual, he sought to modify the historical record in line with his own way of seeing things. In the vision he had in 1503, the celestial voice had encouraged him to think of his discoveries as a direct personal gift of God, which was his to dispose of as he wished. That conviction came back to him now.

I served the King and Queen, our Lord and Lady [he wrote] in the Indies—I say served, but it rather seems that by God's will I gave them to them, as a chattel of my own, I may say, because I had to importune their Highnesses about them, for they were unknown and the way to find them was hidden to all who were asked about them.

He took the opportunity to honour some outstanding obligations of conscience—to creditors, to Beatriz Enríquez, mother of his son Fernando, and to the memory of his father and mother and his wife, for whom, together with his own soul 'and those of all the faithful departed', masses were to be endowed by his son Diego 'when he shall have sufficient income from his said entail and inheritance', preferably in Hispaniola, 'which God gave me by a miracle'. There was still, however, no entirely altruistic bequest—no ecclesiastical foundation that was not to be self-glorifying, no charitable endowment that was not to be primarily for the benefit of Columbus's heirs. In a further codicil, added later, Columbus recalled obligations from his time in Lisbon, in the late 1470s and early 1480s, to those who befriended him or with whom he had contracted debts: all—save for one unnamed Jew 'who dwelt by the gate of the Jewry'—members of Genoese merchant and banking families.[12] All this seems entirely appropriate in a dying man, whose memory is reaching back selectively over the crises and triumphs of life, as he tries to compose his conscience.

It would be wrong, however, to suppose that Columbus died in a mood of retrospection. As he lay on his deathbed, self-pity and anxiety about the terms of his 'contracts' were mingled with dreams of glory still to come. On Queen Isabella's death, she had been succeeded by her daughter and son-in-law, Juana and Felipe, while Ferdinand remained in Castile as regent. The royal couple were in their Burgundian lordships, and were expected to sail from the Low Countries to Spain. According to Las Casas, Columbus took some comfort from the hope of their arrival, perhaps because he was dissatisfied with Ferdinand's posture, perhaps because he felt that a satisfactory outcome for his case was bound to be suspended during the absence of the heirs. By the time they arrived, on 26 April 1506, he was in the grip of his last illness. The last letter to have survived from his hand is addressed to them:

Most serene and very high and mighty Princes, our Lord King and Lady Queen, I trust your Highnesses will believe me when I say that I never hoped so earnestly for my body's health as when I heard that your Highnesses were to come here across the sea, so that I should be able to come to you and place myself at your service and so that you should see the knowledge and experience I have in navigation. But our Lord has ordained otherwise. I therefore very humbly beseech your Highnesses to reckon me among the number of your royal vassals and servants and to take it as certain that, although this sickness now tries me without mercy, I shall yet be able to serve you with such service as the like of it has never been seen before. The untoward circumstances into which I have been plunged, contrary to all rational expectation, and other adversities, have left me in dire extremity. For this reason I have not gone to meet your Highnesses, nor has my son. I very humbly beseech you to accept my purpose and intent, instead of my presence, as from one who hopes to be restored to his honour and estate, as is promised in writing in the terms of my commissions. May the Holy Trinity keep and increase the high and royal estate of your Highnesses.[13]

Probably within about three weeks of that letter, almost certainly on 20 May 1506, Columbus was dead.

Even then, he did not stop travelling. His bones rattled back and forth across the Caribbean almost as much in death as in life. First interred among the Franciscans of Valladolid, his corpse was transferred in 1509 to the family mausoleum his son had founded in Seville. A change of heart by Don Diego caused them to be removed after his death to the sanctuary of the cathedral of Santo Domingo. In 1795, when Santo Domingo became French, they were transferred again to the decency of Spanish soil in Havana, until the 'liberation' of Cuba in the war of 1898 made it seem proper to send them back to Spain, where they were buried under a suitably pompous monument in Seville Cathedral. With each removal, the possibilities of some mistake multiplied, and Santo Domingo, in particular, remains convinced that the true bones of Columbus lingered there.[14] Lovers of irony like to believe the cicerone's joke that Columbus is 'under the billiard tables' of the Café del Norte in Valladolid.[15] But, if the relics of his body are not in Seville Cathedral, those of his mind certainly are; they can be seen in the form of the surviving books from his library and the notes he scrawled in their margins. Nowhere else are travellers susceptible to the magic of relics more likely to find the spirit of Columbus conjured.

That a weaver's son had died titular Admiral, Viceroy, and Governor; that he should have become the founder of an aristocratic dynasty and have established a claim to fame which has made and kept his name familiar to every educated person in the western world: these are achievements which command the attention of any observer and the respect of most. But it can fairly be objected that Columbus's merits should be judged by his contribution to mankind, not his accomplishments for himself. His contemporaries had mixed views of that contribution. The New World did not shine for all beholders with the glow reflected in Columbus's gaze. For anyone who really wanted to get to Asia, it was a Stone-Age obstacle course. After its discovery in the sixteenth century, the New World tended to drift away again from the Old, developing internal economic systems, 'creole' identities, and finally independent states. When Rousseau totted up the advantages and disadvantages that accrued to mankind from the discovery of America, he concluded that it would have been better if Columbus had shown more restraint. Contemporaries as various as Abbé Raynal and Dr Johnson agreed. The fate of America has remained ever since, in a particular tradition, a paradigm of the despoliation of nature and the corruption of natural man. And if the influence of the Old World on the New was pejorative, that of the New upon the Old was slow to take full effect. Only with the improved communications and mass migrations of the nineteenth century, perhaps only with the transatlantic partnerships of the world wars in the twentieth, did the weight of America wrench the centre of gravity of western civilization away from its European heartlands. The potential of most of the continent is unrealized even today. Five hundred years after the discovery, America's hour has still not come.

Still, the sheer extent of the new lands across the Atlantic, and the large numbers of new peoples brought within the hearing of God's world, left the generation of Las Casas and Fernando Colón in little doubt of the potential importance of the events connected with Columbus's life. By 1552 the historian Francisco López de Gómara could characterize the discovery of the New World as the greatest happening since the incarnation of Christ. Yet the same writer denied that Columbus was truly the discoverer of those lands.[16] This was a representative sentiment. Columbus had complained even in his own lifetime of being 'despoiled of the honour of his discovery' and though he was referring to the stintedness of his acclaim rather than to the elevation of the claims of rivals, it is true that his reputation has since suffered repeatedly from attempts to attribute the discovery of the New World to someone else.

The early history of the controversy was dominated by the legal wrangle between Columbus's heirs and the monarchs of Spain over the non-fulfilment of the royal promises of 1492. Any source of doubt that could be cast on Columbus's claim to have performed his side of the bargain was welcome in the prejudicial atmosphere of the first half of the sixteenth century. It was said, for instance, that the New World had formed part of the domains of King Hesperus or that the credit for the discovery belonged to Martín Pinzón, or that it rested with an 'unknown pilot' who had preceded Columbus to the New World by chance and confided his knowledge to the Genoese when on the point of death. It was this last story which López de Gómara repeated; Las Casas heard it treated as comon knowledge when he was a young man in Hispaniola before 1516; in 1535 Gonzalo Fernández de Oviedo dismissed it as a vulgar rumour; and it has been echoed ever since.[17] Testimony was even procured—almost certainly not without deliberate perjury—to deny the amply verified fact that Columbus had visited the American mainland on his third voyage in 1498. The discoverer's sons, Diego and Fernando, strenuously resisted these allegations. Diego had the testimony of numerous favourable witnesses recorded on his father's side and Fernando wrote extensively in defence of his claims. It must be said that whatever Martín Pinzón's role on the first transatlantic voyage, of which we shall never know the whole truth, he joined the enterprise only at a late stage, when Columbus's plans were already well advanced.

Though Columbus was familiar with many mariners' tales of unknown lands in the west, and recorded some of them along with other evidence in support of his theories, the story of the unknown pilot is unacceptable as it stands: it proceeds from biased sources; it is unwarranted by any contemporary authority; and it relies on the hypothesis of a freak crossing such as is otherwise unrecorded in the latitude on which Columbus sailed (although accidental crossings have happened further south, on routes not known to have been frequented before Columbus's time).[18] The argument that the unknown pilot must have existed because Columbus would not otherwise have known where to go reminds one of Voltaire's ironic case in favour of God: if He did not exist, it would be necessary to invent Him. The unknown pilot is not required, even as a comforting fiction. Columbus had assembled sufficient indicators of lands in the west, according to his own standards, by his own researches, without recourse to secret sources. By his own admission, the materials he collected included seamen's yarns about Atlantic lands, which formed only one flimsy strand in the web of evidence. The 'certainty' he is supposed to have evinced, and which can alone be explained, it is said, by some pre-discovery of America, is, as we have seen, another myth. The presumed mariner cannot have helped very much, since his information was insufficient to preclude Columbus's belief that he had found Asia. The Admiral's doubts on that score, when they arose, were clearly attributable to his own observations.

Columbus's rivals soon multiplied when Vespucci's accounts of his voyages became known. There is no doubt that Vespucci's claim—to be fair to him, one must say rather the claim advanced on Vespucci's behalf under his name—to have visited the continental part of the New World before Columbus is false. It rests on the predated relation of his adventure with Alonso de Hojeda in the Gulf of Paria. Columbus had preceded them in those lands by a year, and Columbus's report had almost certainly been the direct inspiration for Hojeda's attempt. Vespucci's modern admirers place more emphasis on the allegation that the Florentine was first to perceive the true nature of America as a continent separate from the Eurasian landmass. But, as we have argued, Columbus has the prior claim in this respect, too, and if Columbus's opinion wavered, no more was Vespucci's unequivocal or unchanging. He explicitly endorsed Columbus's belief in the proximity of the New World to Asia. Only a year after Columbus's death, Martin Wald-seemüller proposed that the new continent be named in honour of

Amerigo Vespucci whom he proclaimed a geographer equal to Ptolemy: six years later, he retracted the suggestion and restored Columbus to the honour of the discovery. In the mean time, however, the new name had begun to stick. It is pointless to call it a misnomer, but it is important to realize that it has to be justified, if at all, by Vespucci's effectiveness as a publicist, not a discoverer.

The irony is that Vespucci's fame came late—after Columbus's death—and did not affect the close personal relations between the two explorers. Vespucci had been admitted the use of Columbus's books; he witnessed one draft of Columbus's will; he dwelt in Columbus's house during the discoverer's last months in Seville. At the time, Columbus's judgement on him was amply justified: 'Fortune', he wrote, 'has been adverse for him, as for so many others. His labours have not brought him the benefits they deserve.' The image of the two under-appreciated navigators comforting one another in Seville is delicious. It is confirmed by other sources: in 1502 Pictro Rondinelli had reported the same version of Vespucci's fortune, rewarded well short of his deserts. Columbus was so taken with him that he recruited him—with every appearance of satisfaction—to join his lobby at court. Thus the two great rivals for the laurels of the discovery exchanged commiseration and recommendations.[19]

Later writers have added to Columbus's rivals by adducing Norsemen from proto-historical sagas or even Welshmen and Hibernians from remoter and more misty myths. The Welsh discovery, along with those of Scots, Poles, Venetians, *et hoc genus omne*, can be relegated to the realms of fantasy. Of the Irish discovery, it is not possible to say much more than that it was technically feasible. America could have been reached by the navigations of hermits who, from the sixth to the eighth centuries, sought desert islands of the North Atlantic in currachs for devotional reasons; the source in which some of them are recorded—the story of St Brendan's search for the earthly Paradise—is so unusual, by the standards of hagiographical literature of its time, that it is tempting to read it as the account of a real voyage. But everything in the text can be explained without pressing the case for Brendan's landing in the New World. Occam's razor slices America out of the account.[20]

The Vikings cannot be so dismissed: that they should have reached America from their colonies in Greenland is inherently probable and though most of the archaeological 'evidence' summoned in support is highly unimpressive, that from L'Anse aux Meadows near the Belle Isle

Strait in Newfoundland is more or less convincing. The story told in the sagas is riven by inconsistencies but can be synthesized, for what it is worth. In 987 Bjarni Herjolfsson attempted to sail from Iceland to Greenland by a route unknown to any of his crew, got lost, and sighted a previously unknown land. After an interval of fifteen years, Leif Eiriksson, son of the founder of the Scandinavian colony on Greenland, followed up Bjarni's find and sailed down a long coast, to parts of which he gave the names of Helluland, Markland, and Vinland. The saga's description of the last is compatible in every way with northern Newfoundland. A colonizing mission followed under Thorfinn Karlsefni—a trader inspired by Leif's story—but ran foul of the hideous natives, the Skraelingar, and had to be abandoned, at an unknown date early in the eleventh century.[21]

This Norse discovery of America is not to be despised. It may have begun accidentally, but was continued in a genuine explorers' spirit. The Icelanders recorded it after the fashion of their time, and the record was preserved at least until 1347, when the last trading or trapping voyage to Markland is mentioned in the Icelandic Annals. Now that the Vinland Map of Yale University Library, once prematurely hailed as confirmation of the sagas, has been shown to be a forgery, the Vinland tradition cannot be said to have continued thereafter, or to have been transmitted in cartographic form until revived by sixteenth-century antiquarians. The assumption that Columbus must have picked it up on his visit to Iceland in 1477, which some scholars have hardly hesitated to make,[22] is extremely rash. What Columbus did is quite distinct from the achievement of the Norse voyagers and by any rational standard immeasurably more significant. First, unless one credits the assumption that he found something in Iceland, their enterprises were entirely independent of each other. Secondly, the Norse route across the Atlantic was by the way of icy stepping-stones: in effect it linked Iceland and Greenland to Markland, and America to Scandinavia only in an indirect and at best very intermittent fashion, whereas Columbus discovered a direct route of easy access. This does not, of course, detract from the prowess of the Icelandic voyagers, who navigated so far in adverse conditions; but it is an important distinction. Thirdly, the Icelanders were themselves custodians of an impressive but peripheral outpost of the civilization of northern Europe, which was only just becoming a very junior partner in the civilization of Latin Christendom at the time; their contacts with the Skraelingar yielded nothing of cultural significance for the rest of the world; Columbus really did help to put a girdle round the earth, in two distinct respects: he discovered two new routes—transatlantic and trans-Caribbean—which opened the link between the major, formerly sundered civilizations of Mesoamerica and the Mediterranean; and he discovered the trade-wind route that led his successors in the South Atlantic—Europe's great highway in the sixteenth century—to the rest of the world.[23] Finally, Columbus was the first in a continuous tradition of transatlantic navigation which has continued to our time: he is therefore our discoverer of America. The Icelandic voyages detract from that role no more than do the presumed Chinese voyages to America's Pacific Coast, or the discovery made by the first inhabitants of the continent, when they crossed the ice-bound Bering Strait more than twenty thousand years ago.

An alternative argument, still connected with the Vikings, but voiced more often by admirers of Vespucci, is that Columbus's discovery of America was no better than that of the Icelanders, since he came upon the New World quite haphazardly and failed to recognize it correctly; one cannot be said to 'discover' something unless one recognizes it for what it is. It has also been said that neither Columbus nor anyone else up to his time anticipated the existence of a second world landmass and that it is therefore imprecise to speak of the 'discovery' of something which the European mind was not conceptually equipped to comprehend. Rather, the discovery of America happened gradually and cumulatively, as, under the influence of further explorations, men's presuppositions became adapted to the facts. Now it is agreed that one cannot be said to have 'discovered' a thing without recognizing it for what it is. Otherwise the event is a mere accident, which will pass unnoticed unless someone else happens to suggest the identification which the finder failed to make. The penicillin will stay in the crucible until it is washed up; the comet will shoot out of sight. Such was not the case, however, when Columbus stumbled on America.

In the first place, the possibilities of just such a discovery as Columbus made—that of a continent separate from the Eurasian landmass—were seriously debated, actively canvassed, and, in some cases, excitedly anticipated among scholars prior to Columbus's departure. As soon as he returned to report, a considerable number of learned commentators jumped to the conclusion that he had found just such an antipodal world. Columbus himself on his third voyage correctly identified the mainland, which he then discovered for the first time, with this rumoured continent. During the virtual derangement brought on by his subsequent sufferings, he forsook the idea, and even while he still embraced it his opinion of the proximity of his discoveries to Asia was grossly exaggerated, but America did not have to be 'invented';[24] the discourse of the day included suitable terms for describing it and classifying it, and Columbus himself was among the first people to make use of them.

Of course the discovery of America was a process, which began with Columbus but unfolded bit by bit after his time, fitfully, without being fully complete until our own time. There has, after all, been a lot of America to discover. The outline of the coasts of South America was not fully complete until about 1540, and although the Atlantic and Pacific coasts of North America were roughly known by that time, the northern coast remained concealed beneath ice until Amundsen cut his way through it in 1905. On the main point at issue between Columbus and posterity—the relationship of America to Asia—Fernández de Oviedo pointed out in the 1530s that the whole truth was still unknown, and so it remained until the early eighteenth century, when the Bering Strait was explored. Many of the important physical features of the interior were still unknown late in the eighteenth century and unmapped until the early nineteenth; only the advent of aerial mapping in the present century—which did not encompass the last secrets of South America until the 1970s—penetrated the final areas to defy exploration. Long as the process has been, Columbus retains a primordial place as its initiator; and the extent to which he advanced it in the span of his own short career is all the more startling against the backdrop of the process as a whole: after alighting on some islands of the Bahamas, he explored much of the coast of Cuba, Hispaniola, Jamaica, Puerto Rico, the Lesser Antilles as far as Dominica, Trinidad, and the coast of the mainland from the mouth of the Orinoco to the Bay of Honduras.

The last argument against ascribing the discovery to Columbus also raises a conceptual problem. Only from the most crassly Eurocentric perspective, it is said, could one speak of the 'discovery' of land which had been well known to its native peoples for thousands of years. It has even been argued, by a highly creditable scholar, with only the faintest trace of detectable irony, that an American discovery of Europe preceded the European discovery of America, when a Caribbean canoe was misdirected across the Atlantic, and that the knowledge of this was Columbus's 'secret'.[25] Whatever one thinks of this prank, it is hard to deny priority to the American discovery of America. This respectable argument would make 'discovery' an almost useless term, by limiting it to uninhabited lands. It misses the point that discovery is not a matter of being in a place, but of getting to it, of establishing routes of access from somewhere else. The peopling of the New World, which was followed by isolation, was conspicuously not a discovery in that sense. So vast a hemisphere naturally afforded scope for much internal exploration: it is proper to speak of exploration, recorded in maps, by some Eskimo, North American Indian, and Mesoamerican peoples, and by the Incas, recorded in the latter case with mnemonic devices which we now barely understand.[26] The reliance of early Spanish and Portuguese explorers on native guides, even in some cases over long distances, suggests that other histories of exploration happened in the New World, which we can only guess at. None of this makes the creation of routes which no one knew about before, such as Columbus's across the Atlantic, any less of a discovery.

Despite nearly five hundred years of assiduous detraction, his prior role in the discovery of America remains the strongest part of Columbus's credentials as an explorer. But we should recall some of the supporting evidence too: his decoding of the Atlantic wind system; his discovery of magnetic variation in the Western hemisphere; his contributions to the mapping of the Atlantic and the New World; his epic crossings of the Caribbean; his demonstration of the continental nature of parts of South and Central America; his *aperçu* about the imperfect sphericity of the globe; his uncanny intuitive skill in navigation. Any of these would qualify an explorer for enduring fame; together they constitute an unequalled record of achievement.

Columbus was a self-avowed ignoramus who challenged the received wisdom of his day. His servility before old texts, combined with his paradoxical delight whenever he was able to correct them from experience, mark him at once as one of the last torchbearers of medieval cosmography, who carried their lights on the shoulders of their predecessors, and one of the first beacons of the Scientific Revolution, whose glow was kindled from within by their preference for experiment over authority. The same sort of paradox enlivened every aspect of his character. His attraction towards fantasy and wishful thinking was ill accommodated in that hard head, half-full already with a sense of trade and profit. In his dealings with the Crown and his concern for his posterity, his mysticism was tempered by a materialism only slightly less intense—like the rich gurus who are equally familiar nowadays in spiritual retreats and business circles. Though religion was a powerful influence in his life, its effects were strangely limited; his devotional bequests were few; his charity began and almost ended at home. The

Indians he discovered he contemplated with evangelical zeal and treated with callous disregard. He was an inveterate practitioner of deception, a perennial victim of self-delusion, but he was rarely consciously mendacious. In dealing with subordinates, he was calculating and ingenuous by turns. He craved admirers, but could not keep friends. His anxiety for ennoblement, his self-confessed ambition for 'status and wealth', did not prevent him from taking a certain pride in his modest origins and comparing the weaver-Admiral with the shepherd-King. He loved adventure, but could not bear adversity. Most paradoxically of all, beyond the islands and mainlands of the Ocean, Columbus explored involuntarily the marchlands between genius and insanity. Times of stress unhinged—sometimes, perhaps, actually deranged—him; in his last such sickness, he obsessively discarded his own most luminous ideas, and never recovered them.

It probably helped to be a visionary, with a flair for the fantastic, to achieve what he achieved. The task he set himself—to cross the Ocean Sea directly from Europe to Asia—was literally beyond the capacity of any vessel of his day. The task he performed—to cross from Europe to a New World—was beyond the conception of many of his contemporaries. To have accomplished the highly improbable was insufficient for Columbus—he had wanted 'the conquest of what appeared impossible'. He died a magnificent failure: he had not reached the Orient. His failure enshrined what, in the long term, came to seem a greater success: the discovery of America.

One cannot do him justice without making allowances for the weakness that incapacitated him for ill fortune. He was too fearful of failure to face adverse reality—perhaps because he had too much riding on success: not only his personal pride, but also the claims to the material rewards on which his hopes for himself and his heirs rested. It is hard to believe, for instance, that his insistence on the continental nature of Cuba was other than perversely sustained in the face of inner conviction; or that he can really have felt, in his wild and self-contradictory calculations of the longitude of his discoveries, the confidence he claimed. The ambition that drove him was fatal to personal happiness. Almost anyone, it might be thought, would rest content with so much fame, so much wealth, so many discoveries, so dramatic a social rise. But not Columbus. His sights were always fixed on unmade discoveries, unfinished initiatives, imperfect gains, and frustrated crusades. Instead of being satisfied with his achievements he was outraged by his wrongs. Unassuaged by acclaim, he was embittered by calumnies. This implacable character made him live strenuously and die miserably. Without it, he might have accomplished nothing; because of it, he could never rest on his laurels or enjoy his success. It was typical of him to abjure his achievement in discovering a new continent because he could not face failure in the attempt to reach an old one. He wanted to repeat his boast, 'When I set out upon this enterprise, they all said it was impossible', without having to admit that 'they' were right.

The Oxford Union Society once invited an American ambassador to debate the motion, 'This House Believes that Columbus Went Too Far'. The eighteenth-century debate on the moral benefits of the discovery of America no longer commands much interest, but we can still ask the less solemn question, 'What difference did it make?' The brouhaha of the fifth centenary celebrations creates the impression of a generalized and unthinking acceptance that Columbus was the prot-agonist of an important event; yet it may still be worth asking what exactly makes it important and what, if any, is the justification for the fuss.

One of the most conspicuous changes to have overtaken the civiliza-tion in which we live—we usually call it 'Western civilization' or 'Western society'—in the course of its history has been the westward displacement of its centre of gravity, as its main axis of communication, the Mediterranean 'frog-pond' of Socrates has been replaced by an Atlantic 'lake' across which we traffic in goods and ideas and around which we huddle for our defence. The career of Columbus, which began in the Mediterranean and took Mediterranean mariners and colonists across the Atlantic for the first time, seems to encapsulate the very change which it can be said to have initiated. At present—and for as long as quincentennial euphoria lasts—the Admiral of the Ocean Sea is bound to seem significant for us. Historians and journalists will even acknowledge, without embarrassment, that he made the sort of personal contribution to history which, in our awareness of the determining influence of the long and grinding 'structures' of economic change, we have become loath to concede to individuals. On the other hand, the judgements of history are notoriously fickle, and depend on the per-spective of the time in which they are made. It may not be long now before 'Western civilization' is regarded as definitively wound up—not cataclysmically exploded, as some of our doom-fraught oracles have foretold, but merely blended into the new 'global civilization' which, with a heavy debt to the Western world but a genuinely distinct identity, seems to be taking shape around us. At the same time, the motors of the world economy are moving or have moved to Japan and California. The Pacific is likely to play in the history of 'global civilization' the same sort of unifying role which the Atlantic has played in that of the West. By 2020, when we come to celebrate the five hundredth anniversary of Magellan's crossing of the Pacific, those of us who are still alive may look back wistfully to 1992 with a feeling of *déjà vu*, and irresistible misgivings about the fuss.

NOTES

Abbreviations Used in the Notes

Bernáldez   A. Bernáldez, *Memorias del reinado de los Reyes Católicos*, ed. J. de Mata Carriazo (Madrid, 1962).

Buron   *Ymago Mundi de Pierre d'Ailly*, ed. E. Buron, 3 vols. (Paris, 1930).

Cartas   *Cartas de particulares de Colón y relaciones coetáneas*, ed. J. Gil and C. Varela (Madrid, 1984).

Décadas   Pedro Mártir de Anglería, *Décadas*, ed. E. O'Gorman, 2 vols. (Mexico City, 1964). (As there are numerous translations and editions of this work, none pre-eminent, I cite it throughout by numbers of decade, book, and chapter; the text I have used is that of the O'Gorman translation.)

Epistolario   Pedro Mártir de Anglería, *Epistolario*, ed. J. López de Toro, 2 vols. (Madrid, 1953) (Documentos inéditos para la historia de España, 11–12).

Historie   *Le Historie della vita e dei fatti di Cristoforo Colombo per D. Fernando Colombo suo figlio*, ed. R. Caddeo, 2 vols. (Milan, 1958).

Las Casas   B. de Las Casas, *Historia de las Indias*, ed. A. Millares Carló, 3 vols. (Mexico City–Buenos Aires, 1951).

Morison   S. E. Morison, *Admiral of the Ocean Sea*, 2 vols. (Boston, Mass., 1942).

Navarrete   *Obras de Martín Fernández de Navarrete*, ed. C. Seco Serrano, i–ii (Madrid, 1954–5). (I have used this edition because I have had it to hand, but the original edition, *Colección de los viages y descubrimientos que hicieron los espãnoles por mar desde fines del siglo XV*, 5 vols. (Madrid, 1825–37), i–iii, is generally preferred.)

Oviedo   G. Fernández de Oviedo y Valdés, *Historia general y natural de las Indias*, ed. J. Pérez de Tudela Bueso, 5 vols. (Madrid, 1959), i.

Pleitos   *Pleitos colombinos*, ed. A. Muro Orejón et al. (Seville, 1964–), i–iv (1964–89), viii (1964).

Raccolta   *Raccolta di documenti e studi pubblicati della Reale Commissione Colombiana*, 6 parts in 14 vols., ed. C. de Lollis et al. (Rome, 1892–6). Parts and volumes are cited by small capital and lower-case roman numerals.

Textos   *Cristóbal Colón: Textos y documentos completos*, ed. C. Varela (Madrid, 1984). Writings of Columbus are normally cited from this work, unless the text seems more satisfactory in another edition.

Thacher   J. B. Thacher, *Christopher Columbus: His Life, Work and Remains*, 3 vols. (New York, 1903–4).

Preface

1. *Textos*, 101.
2. Las Casas, i. 189; this vignette is convincing but unverifiable.
3. *Textos*, 268.
4. See pp. 3, 197 n. 8.
5. Only J. Heers, *Christophe Colomb* (Paris, 1981) has, among modern works, a truly robust attitude to the euhemerism of the 16th-century writings, but, like the other major scholarly biographical studies by S. E. Morison, *Admiral of the Ocean Sea*, and C. Verlinden, *Cristóbal Colón ye el descubrimiento de América* (Madrid, 1967), takes curiously little interest in depicting what Columbus was like. See Morison's defence of reliance on what he calls 'contemporaries', i. 67–8. A monograph which vividly captures some of Columbus's most intimate concerns is A. Milhou, *Colón y su mentalidad mesiánica en el ambiente franciscanista español* (Valladolid, 1983). C. de Lollis, *Cristoforo Colombo nella legenda e nella storia* (Rome, 1892) and J. B. Thacher, *Christopher Columbus: His Life, Work and Remains*, 3 vols. (New York, 1903–4) can still be conscientiously recommended.
6. A. Ballesteros y Berett, *Cristóbal Colón y el descubrimiento de América*, 2 vols. (Barcelona, 1945), i. 90–130, reviews all but one of these theories, with earlier ones—excusable because they predated definitive proof of Genoese provenance—depicting Columbus as Greek, Corsican, English, French, or Swiss. E. Bayerri y Bertomeu, *Colón tal cual fue* (Barcelona, 1961), 451–81, devotes much futile ingenuity to the reconstruction of a Catalan Columbus from the 'Isla de Génova' in Tortosa, a theory which does not require a specific refutation, *a fortiori*. The same may be said of L. Saladini's relatively uninventive recent effort to resurrect the Corsican Columbus (*Les Origines de Christophe Colomb*, Bastia, 1983). The Ibizan Columbus is a relative newcomer, invented by a local journalist who has recruited rash approbation from the present head of the Colón family, the Duke of Veragua.
7. *Textos*, 203, 258; *Raccolta*, I. ii. 366.
8. The most thorough expositions are those of S. de Madariaga, *Christopher Columbus* (London, 1949), esp. 50–65, and S. Wiesenthal, *The Secret Mission of Christopher Columbus* (New York, 1979).
9. E. Vignaud, *Histoire critique de la grande entreprise de Christophe Colomb*, 2 vols. (Paris, 1911), *Le Vrai Christophe Colomb et la légende* (Paris, 1921); J. Manzano y Manzano, *Colón y su secreto: El predescubrimiento* (Madrid, 1982); L. Ulloa, *El predescubrimiento hispano-catalán de América* (Paris, 1928); J. Pérez de Tudela y Bueso, *Mirabilis in Altis: Estudio crítico sobre el origen y significado del proyecto descubridor de Cristóbal Colón* (Madrid, 1983).
10. G. Granzotto, *Christopher Columbus: The Dream and the Obsession* (London, 1986), esp. 121; P. E. Taviani, *Christopher Columbus: The Grand Design* (London, 1986), esp. 86, 109; Madariaga, *Christopher Columbus*, 55, 69.
11. Madariaga, *Christopher Columbus*, 36; Granzotto, *Christopher Columbus*,

121; even Morison succumbs to this temptation, imagining, for instance, Columbus's chats with his wife: i. 161; cf. i. 135.

12. Its publication in 1571 was intended, in any event, to serve the interests of the Colón family. See A. Cioranescu, *La primera biografía de Cristóbal Colón* (Santa Cruz de Tenerife, 1960) and A. Rumeu de Armas, *Hernando Colón: Historiador del descubrimiento de América* (Madrid, 1973).

13. J. Larner, 'The Certainty of Columbus', *History*, 73 (1988), 3–23; Las Casas, i. 72.

14. While this book was in progress, Professor A. Rumeu de Armas brought to my attention a manuscript which has recently come to light, without any convincing history or provenance, purporting to be an 18th-century copy of some Columbus documents, including a number unknown from other sources. I have the most enormous respect for Professor Rumeu's judgement but have felt obliged to leave this manuscript out of account and shall proceed to a detailed critique of it in due course. A facsimile edition is available: *Libro copiador de Don Cristóbal Colón* (Madrid, 1990).

Chapter One

1. *Raccolta*, II. i. 16; *Cristoforo Colombo: Documenti e prove della sua appartenenza a Genova* (Genoa, 1931), 116–17.

2. *Historie*, i. 43–55.

3. Ibid. 55.

4. *Textos*, 189.

5. *Cartas*, 289, 319; Las Casas, i. 497; *Textos*, 339, 351.

6. *Raccolta*, II. i. 84–160.

7. Thacher, i. 190; cf. the emphasis of the early Genoese prosopographers on his humble origins: ibid. 196–207.

8. Las Casas, i. 409; Bernáldez, 333.

9. *Textos*, 269, 272.

10. Las Casas, i. 189.

11. *Textos*, 329, 361.

12. Ibid. 191.

13. See n. 9 above.

14. *Textos*, 166–7.

15. F. Fernández-Armesto, *The Canary Islands after the Conquest* (Oxford, 1982), 136–40; *Monumenta henricina*, 15 vols. (Coimbra, 1960), ix. 55, 129; xi. 110, 142; P. Margry, *La Conquête et les conquérants des îles Canaries* (Paris, 1886), 253; *Fontes Rerum Canariarum* (La Laguna, 1933–), ix. 31, 33; xi. 107, 215; *Historie*, i. 61–2.

16. G. Díez de Games, *El vitorial*, ed. J. de Mata Carriazo (Madrid, 1940), 40–7, 86–96, 201, 256–61, 300; G. Vicente, *Obras completas*, ed. A. J. da Costa Pimpão (Barcelos, 1956), 55.

17. *Textos*, 277.

18. *Historie*, i. 56; Las Casas, i. 31; *Raccolta*, II. iii. 29.

19. Bernáldez, 269–70.
20. *Textos*, 277.
21. Ibid. 89.
22. *Cristoforo Colombo: Documenti e prove*, 187.
23. *Textos*, 306.
24. Ibid. 55, 78.
25. Ibid. 167; *Raccolta*, I. ii. 291, 364–9, 375, 390, 406–7; *Historie*, i. 64–7, indirectly confirmed by *Textos*, 19.
26. S. Münster, *Cosmographia* (Basle, 1554), 139, 178; (1572), 178, 248.
27. *Poesie*, ed. L. Cocito (Rome, 1970), 566.
28. J. Heers, 'Portugais et génois au xve siècle: La Rivalité Atlantique–Méditerrannée', *Actas do III colóquio internacional de estudos luso-brasileiros*, ii (Lisbon, 1960), 141–7.
29. J. Heers, *Gênes au XVe siècle* (Paris, 1961), 200–4, 544–9; M. Balard, *La Romanie génoise*, 2 vols. (Genoa, 1978), ii. 522–31.
30. *Textos*, 188, 363; R. Pike, *Enterprise and Adventure* (New York, 1966), 99, 186, 192–3. Columbus claimed kinship with the Fieschi (*Textos*, 332), who were related to the Centurione, but so far no corroborative documents have come to light.
31. M. Lombard, 'Kaffa et la fin du route mongole', *Annales*, 5 (1950), 100–3.
32. A. Boscolo, 'Gli Insediamenti genovesi nel sud della Spagna all'epoca di Cristoforo Colombo', *Saggi di storia mediterranea tra il XIV e XV scoli* (Rome, 1981), 174–7; F. Melis, 'Málaga nel sistema economico del XIV e XV secoli', *Economia e storia*, 3 (1956), 19–59, 139–63; J. Heers, 'Le Royaume de Grenade et la politique marchande de Gênes en Occident', *Le Moyen Âge*, 63 (1957), 87–121.
33. See pp. 84, 86–9, 106–8.
34. C. Verlinden, *Les Origines de la civilisation atlantique* (Paris, 1966), 167–70.
35. *Textos*, 56.
36. Heers, *Gênes au XVe siècle*, 35–46.
37. Ibid. 271–9.
38. H. Sancho de Sopranis, *Los genoveses en Cádiz antes de 1600* (Larache, 1939); 'Los genoveses en la región gaditano-xericense de 1460 a 1800 [*sic* for 1500]', *Hispania*, 8 (1948), 355–402.
39. M. Ladero Quesada, 'Los genoveses en Sevilla y su región (siglos XIII–XVI): elementos de permanencia y arraigo', *Los mudéjares de Castilla y otros estudios de historia medieval andaluza* (Granada, 1989), 283–312; Pike, *Enterprise and Adventure*, 1–19, 37–9.
40. *Textos*, 314; *Cancionero de Juan Alfonso de Baena*, ed. J. M. de Azaceta, 3 vols. (Madrid, 1956), ii. 497–514; iii. 1105.
41. *Textos*, 167.
42. Ibid. 306.
43. See pp. 79, 172.
44. *Historie*, i. 61.
45. *Textos*, 272.
46. *Cartas*, 205; *Pleitos*, iv. 245.

47. *Cristoforo Colombo: Documenti e prove*, 137.
48. *Textos*, 363.
49. Ibid. 167.
50. *Historie*, i. 64–7, 74–8; Las Casas, i. 66–9.
51. *Textos*, 1–2, 167; Buron, i. 345, ii. 531.
52. J. Gil, 'La visión de las Indias', in *Textos*, pp. xxxvi–xl.
53. A. de Cortesão, 'The North Atlantic Nautical Chart of 1424', *Imago Mundi*, 10 (1953), 1–13; J. de Lisboa, *Livro da marinharia*, ed. B. Rebelo (Lisbon, 1903), 121–2; E. Benito Ruano, *San Borondón, octava isla canaria* (Valladolid, 1978).
54. H. Yule Oldham, 'A Pre-Columbian Discovery of America', *Geographical Journal*, 5 (1895), 221–39.
55. J. A. Williamson, *The Cabot Voyages* (Cambridge, 1962), 197–203.
56. S. E. Morison, *The Portuguese Voyages to America* (Cambridge, Mass., 1940), 32; J. Martins da Silva Marques, *Descobrimentos portugueses*, 3 vols. (Coimbra, 1940–71), iii. 124, 130, 278, 317, 320–32, 552.

Chapter Two

1. 'Andando más, más se sabe'. *Textos*, 218.
2. Ibid. 277.
3. Ibid. 325.
4. See no. 30 below.
5. *Textos*, 203.
6. C. Varela in *Textos*, 287 n. 1.
7. *Textos*, 217.
8. *Cartas*, 267.
9. *Textos*, 277.
10. Ibid. 101, 197, 303; *Raccolta*, i. ii. 75–160; Navarrete, i. 222.
11. F. Fernández-Armesto, 'Atlantic Exploration before Columbus: The Evidence of Maps', *Renaissance and Modern Studies*, 30 (1986), 12–34.
12. E. H. Bunbury, *A History of Ancient Geography*, 2 vols. (London, 1879), i. 620–66. Columbus knew of, but ignored, Eratosthenes. Buron, i. 189.
13. See W. G. L. Randles, 'Le Nouveau Monde, l'autre monde et la pluralité des mondes', *Actas do III colóquio internacional de estudos luso-brasileiros*, iv (Lisbon, 1961), 347–82; Buron, i. 199.
14. D. Bennett Durand, *The Vienna-Klosterneuburg Map Corpus of the Fifteenth Century* (Leiden, 1952), pls. XIII, XV, XVI; J. Parker, 'A Fragment of a Fifteenth-century Planisphere in the James Ford Bell Collection', *Imago Mundi*, 19 (1965), 106–7.
15. H. L. Jones (ed.), *The Geography of Strabo*, i. 243 (I. iv. 6); A. Diller, *The Textual Tradition of Strabo's Geography* (Amsterdam, 1975), 97–134; *Historie*, i. 88; Las Casas, i. 154.
16. St Augustine, *De Civitate Dei*, 16. 9; Macrobius, *Commentaria in Somnium Scipionis*, ed. F. Eyssenhardt (Leipzig, 1893), 614–16 (ii. 8); J. K. Wright,

The Geographical Lore of the Time of the Crusades (New York, 1925), 11, 159–61.

17. *Raccolta*, I. ii. 366–406; Buron, i. 207–15, 235, ii. 427; Aristotle, *De Caelo*, ed. W. H. K. Guthrie (London, 1939), 253.

18. The elaborate case of H. Vignaud, *Toscanelli and Columbus* (London, 1902) has been undermined many times, most effectively by D. L. Molinari, 'La empresa colombina y el descubrimiento', in R. Levene (ed.), *Historia de la nación argentina*, ii (Buenos Aires, 1939), 286–337, but doubts have been revived by A. Cioranescu, 'Portugal y las cartas de Toscanelli', *Estudios americanos*, 14 (1957), 1–17. As all arguments against the authenticity of the letters involve fantastic ramifications, it is tempting simply to wield Occam's razor on their behalf. The main objection—that Columbus never cited Toscanelli among his authorities—is puzzling, but not insuperable: he only developed the habit of citing authorities from 1498, by which time he had plenty of more respectable sources—classical, biblical or apocryphal, and patristic—in his locker. Rumeu de Armas, *Hernando Colón*, 257–88 presents the best case yet in favour of the letters.

19. *Raccolta*, I. ii. 364; I. iii. 67; v. i. 554–88; *Historie*, i. 55–63; Las Casas, i. 62–6.

20. Molinari, 'La empresa colombina', 320–37.

21. *Textos*, 217.

22. The problem of explaining the small differences in the hands which annotated the books has never been satisfactorily solved. Many scholars have contented themselves with ascribing to Columbus all annotations marked with a cross. Las Casas was convinced that Bartolomé had written many of the annotations, including the one in question here (i. 146), but Columbus is more likely to have been in Lisbon at the relevant time than his brother. Some palaeographers have presumed to detect more than two hands: Fernando Colón, for instance, added notes of his own to his father's copy of Marco Polo (J. Gil in *El libro de Marco Polo anotado por Cristóbal Colón*, Madrid, 1987, p. ix) and other discrepancies might be accounted for by the intervention of a fourth hand, such as that of Fray Gaspar de Gorricio, custodian of Columbus's papers (see above, pp. 157–9) or, more convincingly in many cases, by the evolution of the brothers' hands over time. The case for treating as holograph all the notes to d'Ailly and, by implication, to Pius II, is admirably summarized by J. Gil in his contribution to the introduction to *Textos*, pp. lvi–lxii. The note in question here (Buron, i. 207) is reproduced in facsimile in *Raccolta*, I. iii, no. 23.

23. *Raccolta*, I. ii. 407 (no. 490); Buron, ii. 530.

24. Buron, i. 159, 223, 225–7; ii. 522; A. de Altolaguirre y Duvale, *Cristóbal Colón y Pablo del Pozzo Toscanelli* (Madrid, 1903), 4?–3, 375; G. E. Nunn, *Geographical Conceptions of Columbus* (New York, 1924), 1–30; Morison, i. 54–5.

25. The argument in favour of 1485 rests on the assumption that Columbus accompanied in person José Vizinho's expedition of that year, referred to in

inconclusive terms in one of his annotations. *Raccolta*, I. ii, no. 860.

26. *Textos*, 15, 22, 27, 32, 42–5, 48–9, 55, 58, 65, 73–5, 78, 95, 99, 132, 142, 144.

27. Ibid. 24.

28. Las Casas, i. 150.

29. *Epistolario*, i. 307; Navarrete, i. 360–2.

30. Las Casas, i. 155; Bernáldez, 269.

31. Especially by Pérez de Tudela y Bueso, *Mirabilis in Altis*; P. Moffitt Watts, 'Prophecy and Discovery: On the Spiritual Origins of Christopher Columbus's Enterprise of the Indies', *American Historical Review*, 90 (1985), 73–102; and J. Gil, *El Libro de Marco Polo* (Madrid, 1986) but not in the 1987 work of similar title. The annotations are collected and reproduced in facsimile in *Raccolta*. I, ii. 289–525; I. iii (1892) and I. iii (*Supplemento*) (1894). Editions which include the texts exist of Pierre d'Ailly in Buron (but this is incomplete) and, in photostat, in *Imago Mundi by Petrus de Aliaco (Pierre d'Ailly) with annotations by Christopher Columbus* (Boston, 1927) and of Marco Polo in Gil, *El libro de Marco Polo*. I have not seen the edition by L. Giovanni, *Il Milione con le postille di Cristoforo Colombo* (Rome, 1985). Annotations to Pius II are studied in A. Phillimore's unpublished London University M.Phil. thesis, 'The *Postille* of Christopher Columbus to the *Historia Rerum Ubique Gestarum* of Pius II' (1988). I am grateful to Lady Phillimore for lending me a copy of this work.

32. *Textos*, 203.

33. The edition known to Columbus is available in facsimile with an introduction by R. A. Skelton (Amsterdam, 1966). The only good modern edition is *Ptolemaei Geographia*, ed. C. F. A. Nobbe (Leipzig, 1843–5). There is a useful translation, *Geography of Claudius Ptolemy*, tr. and ed. E. L. Stevenson (New York, 1932).

34. C. H. Hapgood, *Maps of the Ancient Sea-Kings* (London, 1979), 4–30. The author goes on to develop highly fanciful theories, but the point about the grid seems possible.

35. *Textos*, 319–20.

36. Morison, i. 54–5.

37. Las Casas, i. 150.

38. M. Polo, *The Description of the World*, ed. A. C. Moule and P. Pelliott, (London, 1938), i. 86, 270–1, 328–9, 376, 378, 424; *Mandeville's Travels*, ed. S. M. Letts, 2 vols. (London, 1953), i. pp. xxii–xxv; *Mandeville's Travels*, ed. M. C. Seymour (Oxford, 1967), p. xiv; Gil, *El libro de Marco Polo*, 30.

39. Buron, i. 144, 159–63 and pl. v opp. p. 272; R. Laguarda Trías, *El enigma de las latitudes de Colón* (Valladolid, 1974), 9–10, 16.

40. G. Caraci, 'Quando cominciò Colombo a scrivere le sue postille?' in *Scritti geografici in onore di Carmelo Colamonico* (Naples, 1963), 61–96; *Raccolta*, I. ii. 291, 376–7 (nos. 23, 621, 753); Buron, i. 206–9, iii. 737. *Imago Mundi by Petrus de Aliaco*, 60.

41. The *Historie* claimed that Columbus had a book *De Locis Habitabilibus*, attributed to Julius Capitolinus, which—if it really existed—seems to have

been related to the same theme (*Historie*, i. 67). Bernáldez may have been thinking of the eighth chapter of *Imago Mundi*, 'De quantitate terrae habitabilis', which was particularly heavily annotated by Columbus, or a lost work of Julius Honorius. A. Riese, *Geographi Latini Minores* (Heilbron, 1870), 15–55, prints all surviving works.

42. Caraci, *Scritti geografici*; *Textos*, 115, 325; *Raccolta*, I. ii, nos. 6, 858, 860.
43. *Raccolta*, I. iii, pl. CI.
44. *Raccolta*, I. iii, *Supplemento, passim*, esp. nos. 6, 13, 18, 28, 47, 57, 66, 75, 88, 92, 112, 243, 329.

Chapter Three

1. *Textos*, 203.
2. Ibid. 15–16.
3. Ibid. 272.
4. Ibid. 268, 277, 329, 357.
5. *Cartas*, 142–3.
6. *Textos*, 219, 357.
7. C. Verlinden, 'A Precursor of Columbus: The Fleming Ferdinand van Olmen (1487)', *The Beginnings of Modern Colonization* (Ithaca, NY, 1966), 181–95; Morison, *Portuguese Voyages in America before 1500*, 44–5.
8. Martins da Silva Marques, *Descobrimentos portugueses*, iii. 124, 130, 278, 317, 320–32, 552.
9. Las Casas, i. 149–50.
10. *Textos*, 357.
11. E. Aznar Vallejo, *La integración de las islas Canarias en la corona de Castilla* (Seville-La Laguna, 1984), 23–87; *Documentos canarias en el Registro del Sello* (La Laguna, 1981), 1–30; G. Chil y Naranjo, *Estudios históricos de las islas Canarias*, 3 vols. (Las Palmas, 1876–91), ii. 632–4; A. Rumeu de Armas, 'La reivindicación por la corona de Castilla del derecho sobre las Canarias mayores', *Hidalguía*, 32 (1959), 11; 'Cristobal Colón y Doña Beatriz de Bobadilla', *El Museo Canario*, 20 (1960), part II, 263–7.
12. J. López de Toro, 'La conquista de Gran Canaria en la Cuarta Década de Alonso de Palencia', *Anuario de estudios atlanticos*, 16 (1970), 332.
13. A. Rubió i Lluch, *Documents per l'història de cultura catalana mig-eval*, i (Barcelona, 1908), 52–4; M. Menéndez y Pelayo, *Historia de los heterodoxos españoles*, 7 (1948), 232 ff.; J. Carreras Artau, *Relaciones de Arnau de Vilanova con los reyes de la casa de Aragón* (Barcelona, 1955), 43–50.
14. Milhou, *Colón y su mentalidad mesiánica*, 361–400.
15. Ibid. 391–3; J. Manzano y Manzano, *Cristóbal Colón: Siete años decisivos de su vida* (Madrid, 1964), 198–200.
16. J. M. Cordeiro Sousa, 'La boda de Isabel de Castilla', *Revista de archivos, bibliotecas y museos*, 60 (1954), 35.
17. *Cartas*, 144–5; an appeal to Medina Sidonia is attested only by tradition, which is also responsible for the chronology I follow here. The period of ducal protection could equally well be deferred to after 1488. The evidence

is assembled—and reviewed with a mixture of credulity and ingenuity—by Manzano, *Cristóbal Colón*, 163–225. Medina Sidonia's involvement may have been suggested by confusion with his role as a major investor in Columbus's second Atlantic crossing (Navarrete, i. 352).

18. *Textos*, 339, 344.
19. Ibid. 309, 362.
20. *Cartas*, 239–40; Rumeu de Armas 'Colón y Doña Beatriz', 259–72.
21. *Pleitos*, iii. 390.
22. See Morison, i. 117.
23. *Pleitos*, iii. 390.
24. Manzano, *Cristóbal Colón*, 97–9.
25. *Cartas*, 142–3; this trip to Portugal cannot be regarded as absolutely proved, but is consistent with much circumstantial evidence: see Manzano, *Cristóbal Colón*, 148–62.
26. F. Fernández-Armesto, 'La financiación de la conquista de las islas Canarias durante el reinado de los Reyes Católicos', *Anuario de estudios atlánticos*, 28 (1982), 343–78; Pike, *Enterprise and Adventure*, 99; Berardi can be presumed to have made some contribution by virtue of his membership of Quintanilla's syndicate, but his specific role may have been exaggerated by some historians who make risky inferences from a debt owed by Columbus and recorded in Berardi's will. See Manzano, *Cristóbal Colón*, 321–8; C. Varela, *Cristóbal Colón y los florentinos* (1988), 49–51.
27. G. Fernández de Oviedo, *Libro de cámara del príncipe Don Juan* (Madrid, 1870).
28. Navarrete, i. 309.
29. Fernández de Oviedo, *Libro de cámara*, 38.
30. G. Maura, *El príncipe que murio de amor* (Madrid, 1953), 48; A. and E. A. de la Torre (eds.), *Cuentas de Gonzalo de Baeza* (Madrid, 1955), 208.
31. *Textos*, 345, 352, 358.
32. Manzano, *Cristóbal Colón*, 267–8; Fernández-Armesto, *The Canary Islands after the Conquest*, 22, 72; M. Serrano y Sanz, *Orígenes de la dominación española en América* (Madrid, 1918), pp. ccxviii–ccxxxi.
33. *Textos*, 243; Navarrete, i. 364.
34. A. Rumeu de Armas, *La rábida y el descubrimiento de América* (Madrid, 1968), 33–41. The visit remains possible, but there is no good reason to believe in it.
35. *Pleitos*, iii. 353; iv. 244–6; viii. 258, 300–1, 339–42; Rumeu de Armas, *La Rábida*, 67–84; *Historie*, i. 76–7; Las Casas, i. 68–9.
36. *Epistolario*, i. 242; Las Casas, i. 168. Their role was alleged in the Columbus family's trial with the Crown in 1515: *Pleitos*, ii. 55.
37. See p. 179.
38. *Textos*, 264, 303; R. O. Jones, 'Isabel la Católica y el amor cortés', *Revista de literatura*, 21 (1962); F. Marquez Villanueva, 'Investigaciones sobre Juan Álvarez Gato', *Anales de la Universidad Hispalense*, 17 (1956).
39. Serrano y Sanz, *Orígenes*, pp. xcvii–clxxii.
40. *Textos*, 299, 361; E. Jos, *El plan y la génesis del descubrimiento colombino*

(Valladolid, 1980), 27; M. Andrés Martín, *El dinero de los Reyes Católicos para el descubrimiento de America, financiado por la diócesis de Badajoz* (Madrid, 1987); Duquesa de Berwick y Alba (ed.), *Nuevos autógrafos de Colón y relaciones de ultramar* (Madrid, 1902), 7; cf. Las Casas, i. 171 for the origin of the myth of the jewels.

Chapter Four

1. *Textos*, 15.
2. Navarrete, ii. 302–4, 332–6, 433.
3. J. Martinez-Hidalgo, *Las naves de Colón* (Barcelona, 1969), updated with the results of archaeological investigation of the presumed wreck of the *Santa María* in *A bordo de la 'Santa María'* (Barcelona, 1976), narrows the range of possibilities and makes reasonable suggestions.
4. *Textos*, 27.
5. Ibid. 17–19.
6. Morison, i. 240–63.
7. P. Adam, 'Navigation primitive et navigation astronomique', *VIe colloque internationale d'histoire maritime* (1966), 91–110.
8. See B. Finney, *Hokule'a: The Way to Tahiti* (New York, 1979).
9. *Textos*, 123.
10. Ibid. 21, 24, 26, 27; Las Casas, i. 191; *Cartas*, 267; *Raccolta*, v. i. 577–80. The elaborate arguments devised against accepting Toscanelli's map as graduated (e.g. S. Crinò, *Come fu scoperta l'America*, Milan, 1943, 59–162) disregard the plain and obvious meaning of Toscanelli's description: *Cartas*, 137–8.
11. Las Casas, i. 189.
12. See p. 139.
13. R. Laguarda Trías, *El enigma de las latitudes de Colón* (Valladolid, 1974), 13–17, 27–8; Buron, i. 144–5, 159–63, pl. v., opp. p. 272.
14. *Textos*, 26.
15. A. Magnaghi, 'Incertezze e contrasti delle fonti tradizionale sulle osservazioni attribute a Cristoforo Colombo intorno ai fenomeni della declinazione magnetica', *Bolettino della Società Geografica Italiana*, 69 (1937), 595–641; Laguarda, *El enigma*, 24–7.
16. *Textos*, 23–4. For my translation of 'alcatraz' as 'heron', see the justification in my *Columbus on Himself* (forthcoming, London, 1992).
17. *Textos*, 21, 206, 211–12. I rely on the assumption that Castilian efforts to locate the line of demarcation between Castilian and Portuguese zones of navigation 100 leagues west of the Azores reflect Columbus's view: Navarrete, i. 317.
18. See p. 000.
19. *Textos*, 20; *Pleitos*, iv. 242–8. See F. Morales Padrón, 'Las relaciones entre Colón y Martín Alonso Pinzón', *Revista de Indias*, 21 (1961), 95–105.
20. *Textos*, 23–4.
21. Ibid. 22, 27; C. Sanz, *El gran secreto de la Carta de Colón* (Madrid, 1959), 367;

M. Giménez Fernández, 'América, "Ysla de Canaria por ganar"', *Anuario de estudios atlánticos*, 1 (1955), 309–36; *Raccolta*, III. ii. 3.

22. *Textos*, 27–30; Navarrete, i. 326.

23. The tedious controversy is reviewed in L. De Vorsey and J. Parker, *In the Wake of Columbus* (Detroit, 1985). The highly publicized J. Judge, 'Where Columbus Found the New World', *National Geographic*, 170 (1986), 562–99, is not helpful. On the cartographic evidence, see K. D. Gainer, 'The Cartographic Evidence for the Columbus Landfall', *Terrae incognitae*, 20 (1988), 43–68.

24. *Textos*, 30–43, 46–118; F. Fernández-Armesto, *Before Columbus* (London, 1987), 223–45; Milhou, *Colón y su mentalidad mesiánica*, 102–11.

25. J. Gil in *Textos*, pp. xxxvi–xxxviii.

26. *Textos*, 33–44, 50, 53.

27. Ibid. 54; Milhou, *Colón y su mentalidad mesiánica*, 145–53.

28. *Textos*, 83–4.

29. Ibid. 97–101; *Historie*, i. 136–7; Las Casas, i. 270–85.

30. *Textos*, 109.

31. Ibid. 118–19, 141; *Cartas*, 260.

32. *Historie*, i. 146–7.

33. Las Casas, i. 313.

34. *Textos*, 145–6.

Chapter Five

1. *Epistolario*, i. 242, 244; Navarrete, i. 362.

2. Textos, 148.

3. Navarrete, i. 311, 393; Pliny, *Historia Naturalis*, 2. 189–90.

4. Navarrete, i. 313; *Epistolario*, i. 236–7, 242–6; *Raccolta*, III. ii. 110.

5. See Ch. 4, n. 19; *Raccolta*, III. i. 143, 146–7; 165–6, 169, 193, 196; ii. 1–6.

6. *Cartas*, 145; Rumeu de Armas, *Hernando Colón*, 271–2; *Epistolario*, i. 245; Navarrete, i. 361; Bernáldez, 308–10; *Raccolta*, III. i. 142, 145, 167; D. Ramos, *Las élites andaluzas ante el descubrimiento colombino* (Granada, 1983), 16–63, shows the diversity of response in Spanish sources, while arguing that Columbus's discoveries were largely accepted as Asiatic in Andalusia, until late 1496 or early 1497.

7. *Décadas*, i. 1. 1; cf. the Genoese prosopographer of 1516 in Thacher, i. 96; Navarrete, i. 361–2.

8. *Textos*, 170–6; A. Rumeu de Armas, *Un escrito desconocido de Cristóbal Colón: el memorial de La Mejorada* (Madrid, 1972).

9. Navarrete, i. 363–4.

10. Ibid. 360.

11. Ibid. 394.

12. *Cartas*, 183.

13. See A. Tio, *El doctor Diego Álvarez Chanca* (Barcelona, 1966) and the characterization in *Décadas*, iii. 6. 4.

14. *Cartas*, 159–60.

15. *Textos*, 154. See, e.g., among sources to which Columbus had access, Pliny, *Historia Naturalis*, 3. 21–6 and *Mandeville's Travels*, ed. Seymour, 132, 205, and for the laws governing enslavement of primitives generally A. Rumeu de Armas, *La política indigenista de Isabel la Católica* (Valladolid, 1969) and J. Muldoon, *Popes, Lawyers and Infidels* (Liverpool, 1979), with some corrections in Fernández-Armesto, *Before Columbus*, 232–3.
16. *Cartas*, 259.
17. Las Casas, i. 358–9.
18. *Textos*, 144; Bernáldez, 295.
19. Las Casas, i. 358–9.
20. Ibid. 358; *Textos*, 150.
21. C. O. Sauer, *The Early Spanish Main* (Berkeley, 1966), 72–5.
22. Las Casas, i. 378.
23. *Textos*, 147–62.
24. Ibid. 238.
25. Bernáldez, 322; *Epistolario*, i. 307–8, 318.
26. Bernáldez, 309; *Textos*, 101.
27. *Cartas*, 217–23.
28. Las Casas, i. 390; *Raccolta*, III. ii. 190–1, fo. 59v; *Textos*, 311, 319–20.
29. Las Casas, i. 389; Peter Martyr's doubts emerge from a comparison of *Epistolario*, 318 with *Décadas*, iii. 12.
30. *Cartas*, 224.
31. Ibid. 264–5.
32. *Cartas*, 258; *Textos*, 224.
33. *Cartas*, 172; *Textos*, 150; Oviedo, 11, 13.
34. *Décadas*, iv. 4; Las Casas, i. 416–20.
35. Oviedo, 64–5.
36. *Historie*, 205–6; *Textos*, 362.
37. Columbus endured it twice. See pp. 148–50.

Chapter Six

1. *Textos*, 204.
2. Ibid. 203.
3. Las Casas, i. 486.
4. *Textos*, 179–88.
5. Navarrete, i. 222.
6. *Textos*, 190–9.
7. Milhou, *Colón y su mentalidad mesiánica*, 69.
8. F. Streicher, for example, proposed an entirely secular reading: *Subscripsi Xpophorus Almirante Mayor de las Yndias* ('Signed Christopher Columbus, High Admiral of the Indies'). Cited by R. Caddeo, in *Historie*, ii. 206; P. E. Taviani, *Cristoforo Colombo: La genesi della grande scoperta* (Novara, 1982), 231. This seems no more helpful than the abundant cryptic readings. Madariaga, *Christopher Columbus*, 403–4, 409, 476–7.
9. *Cartas*, 267–9.

10. Buron, i. 208–15; Aristotle, *De Caelo*, ed. Guthrie, 253.
11. Pliny, *Historia Naturalis*, 2. 66–7; *Textos*, 217.
12. *Textos*, 218.
13. Ibid. 189.
14. J. Gil in *Cartas*, 96–7 nn. 147–8.
15. Navarrete, i. 362.
16. *Textos*, 223; Las Casas, i. 500.
17. See, e.g., J. Parker, 'A Fragment of a Fifteenth-century Planisphere in the James Ford Bell Collection', *Imago Mundi*, 19 (1965), 106–7.
18. H. Yule Oldham, 'A Pre-Columbian Discovery of America', *Geographical Journal*, 5 (1895), 221–39.
19. *Textos*, 221; Las Casas, i. 498.
20. J. Martins da Silva Marques, *Descobrimentos portugueses*, iii. 56, 69, 243–4, 500–11, 546–8, 615, 654–5; *Textos*, 222; Las Casas, i. 498.
21. *Textos*, 224–5; Las Casas, ii. 7–9.
22. *Textos*, 206.
23. Las Casas, ii. 9.
24. *Textos*, 208.
25. Ibid. 211.
26. Ibid. 205.
27. Las Casas, ii. 26. I read 'ganarán' ['will gain'] instead of 'ganaron' ['gained'].
28. Ibid. 238.
29. Ibid. 218.
30. Las Casas, ii. 61–3.
31. Ibid. 63.
32. *Textos*, 212–14.
33. Ibid. 214.
34. Buron, i. 197; *Raccolta*, 1. ii. 375; a bulge at the Equator is attributed by Strabo as a theory to Posidonius (*The Fragments*, ed. L. Edelstein and I. G. Kidd, Cambridge, 1972, 69) but Columbus shows no awareness of this.
35. *Textos*, 216.
36. Ibid. 213.
37. Ibid. 245.

Chapter Seven

1. *Textos*, 244.
2. Las Casas, ii. 69; *Cartas*, 273.
3. *Textos*, 256.
4. Las Casas, ii. 257.
5. Navarrete, i. 423. On Columbus's role as a colonizer, J. Pérez de Tudela Bueso, *Las armadas de Indias y los orígenes de la política de colonización* (Madrid, 1956) is an invaluable guide. S. B. Schwartz, *The Iberian Mediterranean and Atlantic Traditions in the Formation of Columbus as a*

Colonizer (Minnesota, 1986) is a summary which locates Columbus in his proper contexts.

6. Navarrete, i. 428.
7. Ibid. 415.
8. Ibid. 406.
9. See Ch. 5, n. 15.
10. *Textos*, 163.
11. Ibid. 145.
12. Ibid. 243–4.
13. Ibid. 358.
14. A. de Herrera, *Historia general de los hechos de los castellanos en las islas y tierrafirme del mar océano*, ed. A. Ballesteros y Beretta and A. Altolaguirre y Duvale (Madrid, 1934), ii.
15. *Textos*, 253; Navarrete, i. 430; Las Casas, ii. 86–90.
16. *Cartas*, 257.
17. E. R. Service, 'The Encomienda in Paraguay', *Hispanic American Historical Review*, 21 (1951), 230–52.
18. Las Casas, ii. 249.
19. *Cartas*, 271–80.
20. *Textos*, 255–9.
21. Ibid. 245, 256.
22. Ibid. 247–9.
23. *Cartas*, 280–1; Navarrete, ii. 17–20, 60.
24. *Textos*, 263, 270; *Historie*, ii. 71.
25. See pp. 91–2.
26. See pp. 167–70.
27. *Don Quixote*, ii. 52.
28. *Textos*, 244, 265.
29. Ibid. 270, 272.
30. Ibid. 265.
31. Ibid. 266.
32. Fernández-Armesto, *The Canary Islands after the Conquest*, 23–30.
33. Ibid. 245.
34. Bernáldez, 335.
35. *Textos*, 264.
36. *Cartas*, 286–90; M. Krása, J. Polišenský, and P. Ratkoš (eds.), *The Voyages of Discovery in the Bratislava Manuscript Lyc. 515/8* (Prague, 1986), 111.
37. Navarrete, i. 222.
38. *Textos*, 264.
39. Ibid. 272.
40. *Cartas*, 288–9.

Chapter Eight

1. *Textos*, 303.
2. Ibid. 277.
3. See p. 50.
4. *Raccolta*, I. ii. 434; Buron, iii, 737.

5. Milhou, *Colón y su mentalidad mesiánica*, 349–400; A. Cioranescu, *Œuvres de Christophe Colomb* (Paris, 1961), 495 n. 14; Columbus attributed the same doctrine to a Genoese envoy at the Spanish court: *Raccolta*, II. ii. 148, 202.
6. See p. 95.
7. *Raccolta*, I, ii. 75–160; iii. pls. XXXIV–CLVIIII.
8. See p. 181.
9. *Textos*, 272, 283.
10. Ibid. 295–301, 305–10.
11. Ibid. 310.
12. Ibid. 305.
13. Ibid. 308–15.
14. *Cartas*, 298–9.
15. *Textos*, 318.
16. Navarrete, i. 223–5.
17. Ibid. 224.
18. *Textos*, 298.
19. Ibid. 303.
20. Ibid. 317.
21. Oviedo, i. 72. This tradition is rendered credible by the evidence of Roldán's death prior to August 1504 disclosed by Cioranescu, *Œuvres de Colomb*, 473 n. 10.
22. *Textos*, 324–5.
23. Ibid. 319. Columbus seems, from his correspondence with Peter Martyr, to have expected to find the Golden Chersonese since his attempt to measure the longitude of Hispaniola on his second voyage: *Epistolario*, i. 261, 307. The language of Columbus's account of his fourth voyage is, however, cautious on the point and leaves open the possibility that another peninsula had to be negotiated before the Chersonese was reached.
24. *Historie*, ii. 93.
25. *Textos*, 318.
26. Ibid. 320.
27. Ibid. 320–1.
28. Ibid. 322–3.
29. W. Christian, *Apparitions in Late Medieval and Renaissance Spain* (Princeton, 1981), 150–87.
30. *Textos*, 323–4.
31. Morison, ii. 386–7.
32. See pp. 100, 110.
33. *Textos*, 332.
34. Ibid. 336.
35. *Cartas*, 340.
36. *Textos*, 329.
37. Ibid. 320, 327; the terms are similar to those in which he wrote to the Pope before departure: ibid. 311.
38. Ibid. 319–20.

39. Ibid. 331.
40. Las Casas, ii. 316.
41. *Décadas*, iii. 4, 16.
42. *Cartas*, 305.

Chapter Nine

1. *Epistolario*, ii. 88.
2. Textos, 341.
3. Ibid. 329, 337, 339, 350.
4. Ibid. 358.
5. Las Casas, ii. 324–7.
6. Ibid. 326.
7. *Textos*, 357.
8. *Raccolta*, I. iii. 159; *Textos*, 289–91.
9. A. Deyermond, *A Literary History of Spain: The Middle Ages* (London, 1971), 195–200.
10. Cf. *Textos*, 329, where Columbus applies the same allusion to himself.
11. Ibid. 360–2.
12. Ibid. 363. Only one bequest in this codicil, to 'Gerónimo del Puerto' of Genoa, may belong to a context different from that of Columbus's Lisbon.
13. Ibid. 358.
14. The case is made by Thacher, iii. 506–13 and A. Pedroso, *Cristobal Colón* (Havana, 1944), 451–71. For a reply on behalf of Seville, which does not quite live up to its title, see B. Cuartero y Huerta, *La prueba plene* (Madrid, 1963).
15. Granzotto, *Christopher Columbus* 283–5.
16. F. López de Gómara, *Historia general de las Indias*, ed. P. Guibelalde and E. M. Aguilera (Barcelona, 1965), 5, 29.
17. Las Casas, i. 70–1; Oviedo, 15–21; see Preface, n. 9.
18. J. Cortesão, *Los portugueses* (Barcelona, 1947), 588–94.
19. *Textos*, 353; Varela, *Colón y los florentinos*, 66–8.
20. S. E. Morison, *The European Discovery of America: The Northern Voyages* (New York, 1971), 13–31, 81–92; G. A. Williams, *Madoc: The Making of a Myth* (Oxford, 1987) is a splendid account of the Welsh 'discovery'.
21. Morison, *European Discovery*, 32–60; G. Jones, *A History of the Vikings* (London, 1968), 295–306.
22. Taviani, *Christopher Columbus*, 93, 352. Like the predecessors whom he cites, Taviani has no evidence, only a 'conviction' about what 'must' have happened.
23. F. Fernández-Armesto (ed.), *The Times Atlas of World Exploration* (forthcoming, London, 1992), ch. 12.
24. E. O'Gorman, *The Invention of America* (Bloomington, 1961).
25. Pérez de Tudela y Bueso, *Mirabilis in Altis*.
26. R. Tom Zuidema, 'Bureaucracy and Systematic Knowledge in Andean Civilization', in G. Collier, R. I. Rosaldo, and J. D. Wirth (eds.), *The Inca and Aztec States* (New York, 1982), 419–58.

INDEX

Note: Christopher Columbus is referred to in sub-entries as CC.

Index compiled by Frank Pert